Cambridge

T0093169

Elements in Flexible and Large-Area Electronics
edited by
Ravinder Dahiya
University of Glasgow
Luigi G. Occhipinti
University of Cambridge

AN ATLAS FOR LARGE-AREA ELECTRONIC SKINS

From Materials to Systems Design

Weidong Yang*
National University of Singapore
Matthew Hon*
National University of Singapore
Haicheng Yao*
National University of Singapore
Benjamin C. K. Tee*
National University of Singapore

*All authors contributed equally to this work

CAMBRIDGE
UNIVERSITY PRESS

CAMBRIDGE
UNIVERSITY PRESS

University Printing House, Cambridge CB2 8BS, United Kingdom

One Liberty Plaza, 20th Floor, New York, NY 10006, USA

477 Williamstown Road, Port Melbourne, VIC 3207, Australia

314–321, 3rd Floor, Plot 3, Splendor Forum, Jasola District Centre,
New Delhi – 110025, India

79 Anson Road, #06–04/06, Singapore 079906

Cambridge University Press is part of the University of Cambridge.

It furthers the University's mission by disseminating knowledge in the pursuit of
education, learning, and research at the highest international levels of excellence.

www.cambridge.org
Information on this title: www.cambridge.org/9781108749244
DOI: 10.1017/9781108782395

© Weidong Yang, Matthew Hon, Haicheng Yao, and Benjamin C. K. Tee 2020

First published 2020

A catalogue record for this publication is available from the British Library.

ISBN 978-1-108-74924-4 Paperback
ISSN 2398-4015 (online)
ISSN 2514-3840 (print)

An Atlas for Large-Area Electronic Skins

From Materials to Systems Design

Elements in Flexible and Large-Area Electronics

DOI: 10.1017/9781108782395
First published online: August 2020

Weidong Yang*
National University of Singapore

Matthew Hon*
National University of Singapore

Haicheng Yao*
National University of Singapore

Benjamin C. K. Tee*
National University of Singapore

*All authors contributed equally to this work

Author for correspondence: Benjamin C. K. Tee, benjamin.tee@nus.edu.sg

Abstract: Electronic skins are critical for many applications in human-machine-environment interactions. Tactile sensitivity over large areas can be especially applied to prosthetics. Moreover, the potential for wearables, interactive surfaces, and humanoid robots has propelled research in this area. In this Element, we provide an account and atlas of the progress in materials and devices for electronic skins, in the context of sensing principles and skin-like features. Additionally, we give an overview of essential electronic circuits and systems used in large-area tactile sensor arrays. Finally, we present the challenges and provide perspectives on future developments.

Keywords: tactile sensor, function materials, circuits, multifunction, electronic skins

ISBNs: 9781108749244 (PB), 9781108782395 (OC)
ISSNs: 2398-4015 (online), 2514-3840 (print)

Contents

1 Introduction

The importance of touch cannot be overstated. As the largest organ in the human body, the skin is distributed over the whole body, taking up 16% of the body's weight, and it performs somatosensory functions via a complex network of specialized mechanoreceptors, neurons, and synapses in the brain. The somatosensory cortex in the brain is responsible for continuously processing massive amounts of complex signals (action potentials) generated by the mechanoreceptors. These receptors can be found in human skin, including muscle, internal organs, and joints [1], [2]. Upon receiving the appropriate stimuli, the mechanoreceptors generate signals that convey information from these receptors to the somatosensory cortex. This process allows humans to rapidly infer size, shape, stiffness, texture, and temperature of objects in contact with the skin surface [3].

Similarly, tactile sensing is one of the most crucial components to allow machines to perceive external stimuli from the environment and make decisions on actuation and motion path planning. As a result, there is an increasing demand for tactile sensors to exhibit high sensitivity in robotics and human–machine interface devices. The demand for such devices is driving innovation in the design of tactile sensor materials [4]–[6]. To mimic human skin functions, it is essential to carefully choose materials or synthesize novel materials that enable optimized sensing performance. Finally, there is a need to integrate tactile sensors with new semiconductors or high-performance circuits and systems for performing the electronic perception. Figure 1.1 illustrates an atlas for materials and systems design of large-area electronic skins. Tactile sensors, as core components of electronic skins, provide new opportunities in developing innovative functional materials and sensitive microstructures, integrating fast and high-efficiency circuits, and finally presenting complex, multi-sensing flexible and stretchable sensor systems.

Numerous tactile sensors based on various sensing principles have been developed. Sensors that incorporate features such as stretchability [22], [23], [24], self-healing [25], [26], self-powered capability [27], [18], solar-powered capability [28], [29], biodegradability [30], [31], light emission [32], [33], and multifunctional sensing [33], [20] have gained popularity in literature. The use of nanostructures in polymers and the patterning of surface microstructures have enabled conformable sensors that can achieve high sensitivity. However, the design of such sensors is still challenging. Despite the challenges, promising applications in electronic skin, smart wearable electronics, and human robotics have continued to propel research and development of novel sensors [34], [35], [36].

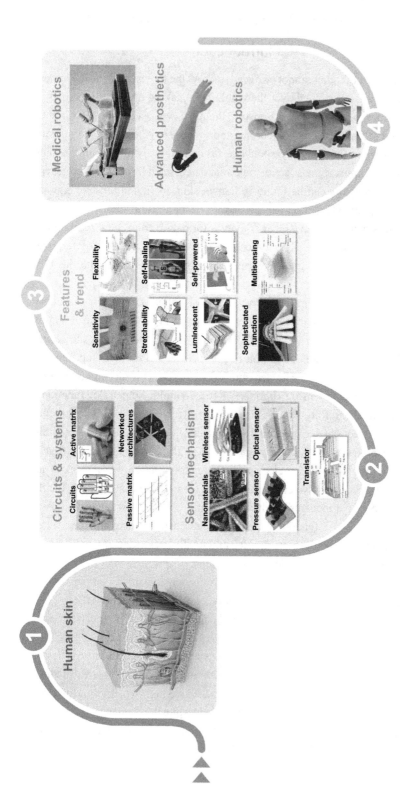

Figure 1.1 An atlas for materials and systems design of large-area electronics skin. Circuits and systems: "Circuits": Reproduced with permission from Ref. [7]. Copyright 2015 American Association for the Advancement of Science. Reproduced with permission from Ref. [8]. Copyright 2005 National Academy of Sciences. Sensor mechanism: "Nanomaterials": Reproduced with permission from Ref. [9].

Caption for Figure 1.1 (cont.)

Copyright 2011 John Wiley and Sons. Reproduced with permission from Ref. [10]. Copyright 2017 John Wiley and Sons. Reproduced with permission from Ref. [7]. Copyright 2015 American Association for the Advancement of Science. Reproduced with permission from Ref. [11]. Copyright 2013 Nature Publishing Group. Reproduced with permission from Ref. [12]. Copyright 2013 Nature Publishing Group. "Flexibility": Features and trend: "Sensitivity": Reproduced with permission from Ref. [13]. Copyright 2018 Nature Publishing Group. "Stretchability": Reproduced with permission from Ref. [14]. Copyright 2018 the American Association for the Advancement of Science. "Self-healing": Reproduced with permission from Ref. [15]. Copyright 2018 American Association for the Advancement of Science. "Luminescent": Reproduced with permission from Ref. [16]. Copyright 2012 Nature Publishing Group. "Self-powered": Reproduced with permission from Ref. [18]. Copyright 2016 John Wiley and Sons. "Sophisticated function": Reproduced with permission from Ref. [19]. Copyright 2016 American Association for the Advancement of Science. "Multi-sensing": Reproduced with permission from Ref. [20]. Copyright 2018 Nature Publishing Group. "Medical robotics": Reproduced with permission from Ref. [21]. Copyright 2018 Springer Nature.

To demonstrate these sensors, electronic hardware is needed to provide signal conditioning, data conversion, and telemetry. The size and distributed natures of large-area conformable sensor arrays make this endeavor especially challenging. In this respect, the choice of the array/network architecture plays an important role in allowing sensors to scale up to cover a large area. The chosen architecture will influence tactile sensor (taxel) density, temporal resolution, coverage area, power, and form factor – all of which are important considerations for a practical implementation. Research in this area has been driven mainly by the engineers and scientists in the robotics community who are tasked with implementing such systems [37]–[39]. The chasm between the development of novel benchtop sensors and the lack of device translation into useful applications has been noted by several researchers [37]. The disconnection between system requirements and sensor development is one reason. Also, the lack of a unified approach and use of off-the-shelf components tend to lead to bulky hardware with lower energy efficiency. Moreover, the emerging Tactile Internet can effectively integrate software and hardware of tactile sensory systems to provide a medium for real-time transmission of control and sensing information, thus enhancing the interaction of humans and machines with their surrounding environment [40].

In this review, we aim to create an updated map or atlas for electronic skins by focusing on the recent progress of materials design and functional integration of large-area flexible tactile sensors. We will cover key topics in sensing mechanisms, material design, sensor electronics, sensor features, and the trend of tactile sensor design in this Element. In Section 2, we highlight the different sensing mechanisms and their corresponding conformable material fabrication strategies. These tactile sensors, based on capacitive, piezoresistive, piezoelectric, triboelectric, optical, and wireless sensing principles, are described with specific functional materials design in the latest literature. In Section 3, we provide an overview of sensor array and network architectures, and briefly discuss some of the interface electronics used for signal conditioning. In Section 4, we give a summary of current breakthroughs in the design of critical features of tactile sensors. We talk about the trends in material design and multifunctional integration. To conclude, we lay out the significant challenges and give perspectives on the future development and exciting potential applications of large-area tactile sensors for electronic skin and wearables.

2 Sensing Mechanism of Tactile Sensors

Tactile sensors are designed to detect applied loads via the transduction of pressure/strain to an electrical output. The various applications of tactile devices also have different requirements for the design and function of touch-sensing elements. Apart from the fundamental properties of pressure sensors such as sensitivity, response time, and reliability, different applications have some special requirements. For example, the tactile sensor on a surgical robotic hand demands highly precise operations, and thus the tactile sensors on robots need high precision in position and force feedback [41]. Actively controlling robot by the human in a human-machine interface is the main task, which transfers human information to the robot or to teach the robot to complete the specified action. In this case, the requirement of operation safety must be satisfied so that the robot is effectively manipulated and meanwhile causes no harm to the human. Moreover, advanced prosthetics such as artificial limbs are useful to help disabled patients grasp or manipulate objects they use in their daily lives. Apart from the safety issue, the applied force of prosthetics should be restricted and stable in order to identify the postures and motions of patients. Therefore, touch sensing is crucial for various tactile sensors and robotics, and tactile function similar to that of human skin is realized by different designs of tactile sensors.

Corresponding to the various types of output signals, sensors can be divided into different types. Typical tactile sensors can be capacitive, piezoresistive, piezoelectric, or optical. Besides, there are other kinds of sensors that are also capable of sensing pressure information (e.g., transistor, and crack-induced sensor). In this section, sensing mechanisms of different sensors will be introduced. Typical examples will be given to show functional materials design of sensors for sensitive pressure response, stretchability, and other performances.

2.1 Capacitive Sensor

Capacitive sensors are already commercially used in various applications (e.g., robotics sensing, human–machine interfaces, and touch screens [45]–[49]). The parallel-plate capacitor is typically designed and studied, which usually consists of two electrodes and dielectric materials. The sensing principle of the capacitive sensor is based on capacitance change due to applied pressure.

$$\mathbf{C} = \varepsilon_0 \varepsilon_r \frac{A}{d}. \tag{1}$$

When pressure is applied, the capacitance increases because of the reduction of spacing, d, between the two electrodes. In this case, the deformation ability of dielectric materials can affect sensitivity in a considerable way. Also, the effective dielectric constant may increase and lead to increase in capacitance during the sensor deformation. An example of a capacitive sensor is given by Bao and coworkers. They developed a sensor by sandwiching a micropyramidal

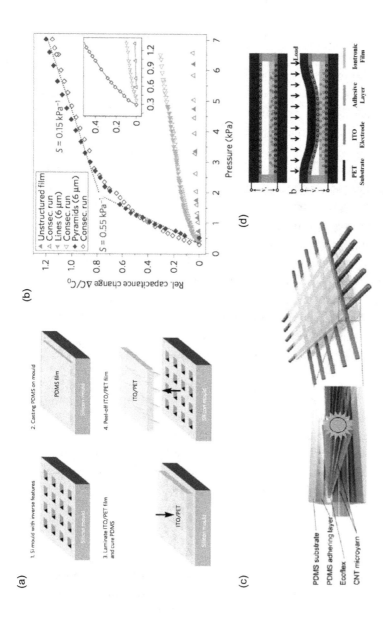

Figure 2.1 Capacitive tactile sensor. (a) The fabrication process of micropyramid-based capacitive sensor. (b) Enhanced sensitivity of microstructures. Reproduced with permission from Ref. [42]. Copyright 2010 Nature Publishing Group. (c) Schematic of sensor consisting of CNT microyarns. Reproduced with permission from Ref. [43]. Copyright 2015 John Wiley and Sons. (d) Schematic showing sensing principle of the ionic capacitive sensor. Reproduced with permission from Ref. [44]. Copyright 2015 John Wiley and Sons.

dielectric material between two PET substrates, which have a conductive layer on the inner surface that acts as the electrode [42]. The sensor fabrication process is shown in Figure 2.1(a). Micropyramidal PDMS was made by casting PDMS on a Si mold. PET with an ITO layer on its surface was selected as the substrate. ITO is a conductive thin layer and can be used as an electrode. The lamination of two PET/ITO layers and micropyramids forms the capacitive sensor. The micropyramids are deformed when pressure is applied, leading to decrease of the spacing and increase in the capacitance between the two electrodes. Also, when the dielectric material is deformed, the effective dielectric constant between the two electrodes also will increase. This also contributes to the increase in capacitance. Figure 2.1(b) shows the capacitance change with the increased pressure. Enhanced sensitivity is exhibited for this micropyramid-based pressure sensor.

Besides, elastic conductors (e.g., carbon nanotube [CNT], silver nanowire [AgNW], and conductive fiber) also can be integrated into the polymer matrix and fabricated as the capacitive sensor. These materials enable the multidirectional sensing of tactile information. For example, Lipomi et al. presented laminated layers of PDMS and Ecoflex, deposited with orientated single-walled carbon nanotubes (SWCNTs), as a capacitive sensor. The sensor is transparent and stretchable, with the ability to detect both normal and tensile forces [50]. Also, hierarchically engineered elastic CNT fabrics were incorporated in the skin-like sensor, as reported in Figure 2.1(c). CNT microyarns were aligned in a point-to-point orientation in Ecoflex to make the pressure sensor [43], achieving good sensitivity and a low detection limit (0.4 Pa). The unique materials alignment within this sensor enables the high spatial resolution. Besides pressure detection, the sensor is also capable of detecting other stimuli (e.g., temperature, humidity, and biological variables), because of the metallic features of CNTs. In addition, stretchable conductors made of AgNWs were also reported in capacitive sensors [51]. Patterned AgNWs were printed and laminated with PDMS and liquid metal layers. This type of sensor is able to detect pressure even up to 1.2 MPa. Another example of elastic conductors used in the capacitive sensor is designed conductive fibers [52]. The absorption of Ag precursors in the SBS layer was reduced to Ag nanoparticles. Such a conductive layer was coated on Kevlar fiber. PDMS was further coated on the surface as the dielectric material. Two composite-coated fibers were laminated to form the sensor.

Another type of capacitive sensor is based on the use of ionic gel. Nie et al. demonstrated a thin-film sensing material by utilizing an ionic gel matrix [44]. Figure 2.1(d) shows the sensor schematic and working principle. Two PET/ITO films were selected as substrates and electrodes, and laminated in a parallel-plate configuration. One of the electrodes was coated with the ionic gel. When the two PET/ITO films were assembled together by an adhesive layer, the ionic gel layer was sandwiched inside. The capacitive sensor was formed, with a gap between the ionic gel and one of the electrode layers. When loads were applied, the electrode made contact with the ionic gel layer, and an electrical double layer (EDL) was formed. Within this EDL, ions in the ionic gel layer were attracted by electrons in the electrode layers at a nanoscale distance. Hence, an ultrahigh unit-area capacitance could be acquired. When the pressure increased, the contact area between the electrode and ionic gel would further increase. The sensor exhibits the sensitive change of capacitance due to the applied pressure.

2.2 Piezoresistive Sensor

The sensing mechanism of the piezoresistive sensor is based on resistance change due to applied pressure. Usually, the contact area between conductive materials and electrodes increases with pressure. Hence, a conductive path is formed and increased, leading to the change in resistance. Two types of piezoresistive sensors are introduced and discussed here. The first one relies on the incorporation of nanomaterials (e.g., carbon nanotube [CNT], graphene, and nanowires). The other one relies on a conductive coating on the surface of polymer matrix, making the whole composite a highly sensitive pressure sensor.

Nanomaterials (e.g., CNT, graphene, and nanowires) show excellent mechanical and electrical properties. Recent research has frequently applied them in the design of sensor materials to develop highly sensitive and flexible pressure sensors. The quantum tunneling effect observed in nanomaterials provides attractive electrical performance in sensor applications. Robust but flexible mechanical features make nanomaterials promising in the development of flexible devices. Figure 2.2(a) gives an SEM image of ZnO nanowires on polystyrene nanofiber (PSNF) and its zoomed-in image [9]. A ZnO/PS hybrid network was formed as ZnO nanowires uniformly grew on the PSNF. The large area of the ZnO nanowires indicates an ample surface contact and hence good conductivity of the sensor when pressure is applied. The flexibility of nanoscale structures promises its application in the flexible pressure sensor. Figure 2.2(b) shows an example of the gold nanowires (AuNWs) used in the wearable sensor [53]. Kimberly-Clark tissue paper was coated with AuNWs and integrated with interdigitated electrodes on the PDMS sheet. When pressure was applied, the deformed conductive nanowires enlarged their contact with electrodes. More conductive paths were formed among the nanowires, and the resistance was

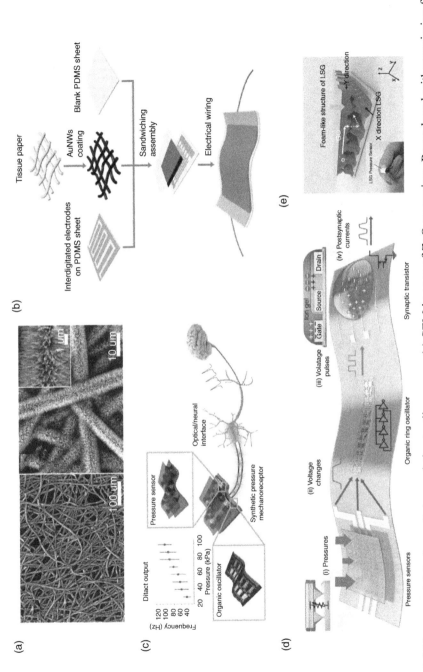

Figure 2.2 Nanomaterial designs of piezoresistive tactile sensors. (a) SEM images of ZnO nanowires. Reproduced with permission from Ref. [9]. Copyright 2011 John Wiley and Sons. (b) Schematic and structure of AuNWs coated wearable sensor. Reproduced with permission

Caption for Figure 2.2 (cont.)

from Ref. [53]. Copyright 2014 Nature Publishing Group. (c) CNT-doped piezoresistive sensors as mechanoreceptors for artificial skin. Reproduced with permission from Ref. [54]. Copyright 2015 American Association for the Advancement of Science. (d) Artificial afferent nerve system based on piezoresistive sensor containing CNT. Reproduced with permission from Ref. [2]. Copyright 2018 American Association for the Advancement of Science. (e) Patterned graphene layers for the sensor. Reproduced with permission from Ref. [55]. Copyright 2015 Nature Publishing Group.

reduced. The porous nanowire network makes the sensor work with both excellent mechanical properties and electrical sensitivity.

Figure 2.2(c) gives an example of CNT used in the piezoresistive sensor. In this work, Tee et al. built up a skin-inspired system to convert detected tactile information to biological signals [54]. To produce a piezoresistor with a wide range of resistance change when responding to increased pressure, the micropyramidal elastomer based on polyurethane and CNT was designed. CNT was dispersed in polyurethane, and the composite was cast to make the micropyramidal sensing elements. Based on the tunneling effect, the conductive path between the CNTs was formed during the deformation of the micropyramids. The concentration of dispersed CNTs and geometrical conditions of the micropyramids could be adjusted for a desirable working range of the sensor. By utilizing a similar sensing mechanism, Bao and coworkers developed an artificial nerve, which is shown in Figure 2.2(d) [2]. CNT-incorporated piezoresistive sensors were made to sense the tactile pressure from multiple fields. Signals were converted to pulse voltages, which were further integrated and processed to actuate the efferent nerve. In addition, multiwalled carbon nanotubes (MWCNTs) were also used in the tactile sensor array. The structural design of sensor materials and coplanar electrodes enabled a cross talk suppression effect for sensor application [56]. Also, a carbon nanotube and silver nanoparticle composite was developed and presented in a whisker-shaped pressure sensor [57]. CNTs in the conductive network matrix showed excellent bendability, while Ag nanoparticles enhanced the sensor conductivity.

Another example shows the use of graphene in the piezoresistive sensor, which is illustrated in Figure 2.2(e) [55]. Foam-like structures of graphene layers were sandwiched together to form a conductive path between the two electrodes. Sensitive pressure responses were achieved due to the double-layer structure of graphene. The laser-scribed design of the graphene layers also enabled a wide sensitivity range in the pressure sensor. The sensor can respond to pressure, even under a large pressure up to 50 kPa. Besides, graphene film can also be further designed for better sensing performance. For example, a graphene woven fabric (GWF) was fabricated with unique polycrystalline structures compared with graphene films [58]. Meshed structures and high density of cracks in the sensor lead to the high gauge factor of this strain sensor. Other designs (e.g., microstructured graphene arrays, and graphene oxide sheet on nanofibers) also have been studied in recent work for highly sensitive pressure sensors [59], [60].

Besides forming the conductive path between electrodes and nanoscale materials, the design of a conductive coating on microstructured elastomers can also make the piezoresistive sensor useful for pressure detection. Choong

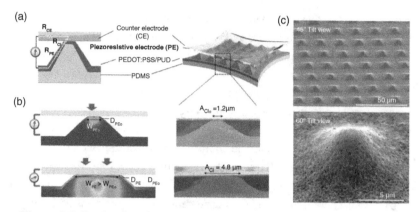

Figure 2.3 Conductive elastomer on the surface of microstructures for the piezoresistive tactile sensor. Reproduced with permission from Ref. [61]. Copyright 2014 John Wiley and Sons.

et al. presented a novel design of sensor by drop casting a conductive PEDOT: PSS layer on the surface of micropyramids made of PDMS [61]. When pressure is applied, the micropyramids are deformed. The resistance will change due to the change of contact area between the micropyramids and the electrodes. This sensing mechanism relies on the geometry deformation of microstructures. A sensitive pressure response, especially at a small pressure, is exhibited due to the small contact area of the pyramid vertex. Figure 2.3(a) illustrates the circuit model to explain the mechanism of this piezoresistive sensor, based on the change in contact area. Figure 2.3(b) shows the simulation results of the deformation of the micropyramids, together with SEM results. When a small pressure was applied, the small contact area between pyramid and electrode generated a large resistance. As the contact area increased with pressure, the resistance decreased. Figure 2.3(c) shows PEDOT:PSS/PUD-coated micropyramids, indicating the good coating resulting from drop casting.

The previously discussed capacitive and piezoresistive pressure sensors are important for touch sensing and are also extensively utilized for detection and control in novel flexible electronics. However, the cross talk issues would cause measurement noise and degrade the sensing performance of the pressure-sensing arrays. The cross talk effects originate from interference between adjacent pressure-sensing elements. To eliminate the cross talk effects, Karmakar et al. fabricated spherical gold nanoparticles (Au-NPs) into the poly (3,4-ethylenedioxythiophene):polystyrene sulfonate (PEDOT:PSS) [62]. The resultant Au-NPs/PEDOT:PSS pressure-sensing arrays without cross talk issues can be used to design high-resolution, multiple-electrode array systems.

2.3 Piezoelectric Sensor

Compared to the general cross talk issues of piezoresistive sensors and the low-pressure sensitivity of capacitive sensors, the piezoelectric sensors possess advantages in greater force sensitivity, multipoint tactile sensing, and spatial resolution. Moreover, since piezoelectric materials and structures can generate electrical signals due to the piezoelectric effect without voltage input, the piezoelectric mechanism has been another good candidate for designing smart tactile sensors. For example, Dagdeviren et al. in Rogers' group reported a conformal device with thin, inorganic, piezoelectric materials based on soft substrates, which enhance piezoelectric response with high sensitivity (~ 0.005 Pa) and fast response time (~ 0.1 ms), respectively. Lead zirconate titanate ($PbZr_{0.52}Ti_{0.48}O_3$, PZT) is one typical piezoelectric material that has been extensively used in microelectromechanical devices and actuators, pressure sensors, because of its high dielectric permittivity, huge piezoelectric coefficients, and remnant polarization. Figure 2.4(a) shows the device schematic, including a piezoelectric film transducer with PZT film, and a cross section of a tactile sensor, which is related to a transistor [63]. The resultant skin-mounted pressure sensor offers good performance in detecting tiny motions on human skin. Although PZT material has a high piezoelectric coefficient, the high processing temperature of PZT (600°C) decreases its range of application due to the fact that most flexible substrates have low thermal processing budgets (< 200°C). Compared to the fabrication process of PZT, ZnO nanowires are easily grown by a hydrothermal process below 90°C, and the piezoelectric coefficient of ZnO nanowire structure is relatively high owing to the structure of the nanowires in the radial direction. Therefore, Jeong et al. [65] utilized ZnO piezoelectric nanowires to design a psychological touch sensor that can induce an electrical "pain" signal to mimic the feeling of human skin pain when a pen cap presses the device, as shown in Figure 2.4(c). This electric "pain" signal is expected to improve the protection mechanism of an android robot or mobile phone against harsh environments.

To date, piezoelectric polymers have attracted intensive research for developing highly sensitive electronic devices because they can sense small forces via pressure, stretching, bending, or twisting. Recently, Persano et al. in Rogers' group also reported employing aligned arrays of nanofibers of poly(vinylidene-fluoride-co-trifluoroethylene) P(VDF-TrFE) to fabricate high-performance piezoelectric sensors [64]. P(VDF-TrFE) generally exhibits good piezo- and ferroelectric behavior and a single all-trans polar crystalline phase (β-phase), which maintain stability at room temperature. The piezoelectricity results from

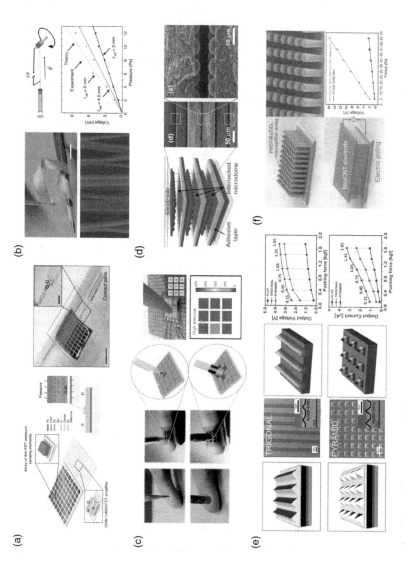

Figure 2.4 Piezoelectric tactile sensors. (a) Illustration of PZT-film piezoelectric pressure sensor and its connections to an associated transistor. Reproduced with permission from Ref. [63]. Copyright 2014 Nature Publishing Group. (b) Highly aligned arrays of oriented

Caption for Figure 2.4 (cont.)

nanofibers of aligned polymer chains of P(VDF-TrFE) and their output voltage under distinct pressure. Reproduced with permission from Ref. [64]. Copyright 2013 Nature Publishing Group. (c) Psychological tactile sensor with electrical "pain" signal based on ZnO piezoelectric nanowires. Reproduced with permission from Ref. [65]. Copyright 2015 Royal Society of Chemistry. (d) Schematic of multilayer interlocked microdome geometry of a (rGO)/PVDF composite film. Reproduced with permission from Ref. [66]. Copyright 2018 American Chemical Society. (e) Microstructured P(VDF-TrFE) piezoelectric devices including output voltage and current and simulations of trigonal line-shaped and pyramid-shaped sensors. Reproduced with permission from Ref. [67]. Copyright 2015 John Wiley and Sons. (f) Hybrid P (VDF-TrFE)/BaTiO$_3$ piezoelectric nanocomposite micropillar arrays and the output voltage of correspondent tactile sensor. Reproduced with permission from Ref. [68]. Copyright 2017 John Wiley and Sons.

electrical dipoles induced by hydrogen and fluorine atoms in the VDF molecules, which are located vertically to the polymer backbone. Figure 2.4(b) depicts the experimental setup and SEM micrograph for electrospinning highly aligned arrays of oriented nanofibers of aligned polymer chains of P(VDF-TrFE), and a comparison of experimental and theoretical pressure response curves at different effect contact length [64]. The resultant pressure sensors indicated good response in a small pressure range, which suggests potential application opportunities in human motion monitoring and robotics.

Although PVDF and its copolymer P(VDF-TrFE) are ideal candidate sensing materials for large-area tactile sensors because of their better mechanical flexibility, processing simplicity, and excellent biocompatibility compared to inorganic piezoelectric materials, the lower piezoelectricity coefficients of PVDF and its copolymers result in significantly lower power outputs than those of inorganic piezoelectric devices. To address this issue, the hybridization of composites of high-piezoelectric inorganic fillers and high-flexibility polymer matrix is one effective approach to impart flexibility and robustness with good performance. For example, Chen et al. recently presented a flexible piezoelectric nanogenerator by hybridizing P(VDF-TrFE) and $BaTiO_3$ to nanocomposite micropillar arrays, as shown in Figure 2.4(f). The piezoelectric device exhibits an enhanced voltage of 13.2 V and a current density of 0.33 $\mu A\ cm^{-2}$, which is 7.3 times that of pristine P(VDF-TrFE) bulk film [68]. On the other hand, the use of microstructured pillar array was attributed to its high sensitivity for detecting human vital signs including different modes of breath and heartbeat pulse, which indicates its potential applications in flexible electronics and the medical sciences. In addition, Lee et al. designed three types of microstructured P(VDF-TrFE) film-based piezoelectric nanogenerators with high power-generating performance for highly sensitive self-powered pressure sensors. Figure 2.4(e) illustrates two nonflat microstructures such as trigonal line-shaped and pyramid-shaped P(VDF-TrFE) piezoelectric nanogenerators and their SEM micrographs, FEM simulation results, and electromechanical coupling comparison [67]. They found that the microstructured P(VDF-TrFE) piezoelectric devices are ultrasensitive to mechanical deformation and present nearly five times large a power output in comparison with the flat film-based device. Moreover, Lee et al. recently reported a flexible ferroelectric sensor with ultrahigh pressure sensitivity over a broad linear range based on multilayer interlocked microdome patterns. Figure 2.4(d) shows the schematic of a multilayer interlocked microstructured e-skin and a cross-section SEM image of the interlocked microdome geometry of a reduced graphene oxide (rGO)/PVDF composite film [66]. Due to the conductive and ferroelectric nature of polymer composites incorporating PVDF and rGO, the resultant e-skins can perceive and distinguish static and dynamic pressure using piezo-resistive and piezoelectric modes with high sensitivity of 47.7 kPa^{-1} and 1.3 Pa minimum detection, respectively. Also, their microstructured piezoelectric sensor is expected to monitor diverse stimuli from a low- to a high-pressure range including weak gas flow, acoustic sound, wrist pulse pressure, and respiration.

Furthermore, the piezoelectric oxide semiconductor field-effect transistor (POSFET) as one special piezoelectric device can measure dynamic contact forces and temperature change. The POSFET as an "integral sensing unit" consists of a P(VDF-TrFE) piezoelectric thin film–based transducer and the related electronic unit. The design process of the POSFET device is divided into two steps: (i) the fabrication of the metal oxide semiconductor (MOS) device, (ii) the deposition and in-situ process of the piezoelectric polymer layer. Dahiya's group has fabricated optimized POSFET devices to detect dynamic normal force over a wide range (0.15–5 N), frequency (2 Hz–2.13 kHz), and average response of about 50 mV/N [41]. Although the piezoelectric polymers show anisotropic behavior, namely, inability to decouple normal force and tangential force, collecting gradient responses of multiple POSFETs in an array can obtain information on the tangential force. Therefore, the multifunctional POSFET with its small size and linear response in a wide range of dynamic forces is suitable for detection of body sites such as fingertips of robotics.

2.4 Triboelectric Sensor

Wang's group first proposed the concept of the triboelectric nanogenerator (TENG), which utilized electrostatic charges generated on the top and bottom faces of two distinct materials. The touch-induced triboelectric charges cause a potential difference, and when the two faces are separated, electrons move between the two electrodes [72].

For instance, Zhu et al. reported a self-powered and flexible tactile sensor based on contact electrification, which induces voltage signals when responding to physical contacts without needing an external power supply [27]. The self-powered tactile sensor consists of a layer of PET as the structural backbone, which is sandwiched by transparent ITO layers as electrodes on both sides. On the top side, a layer of fluorinated ethylene propylene (FEP) as an electrification layer can generate triboelectric charges while in contact with another object, and surface modification on the FEP is used to fabricate vertically aligned polymer nanowires, which is of importance in obtaining high sensitivity for low pressure detection.

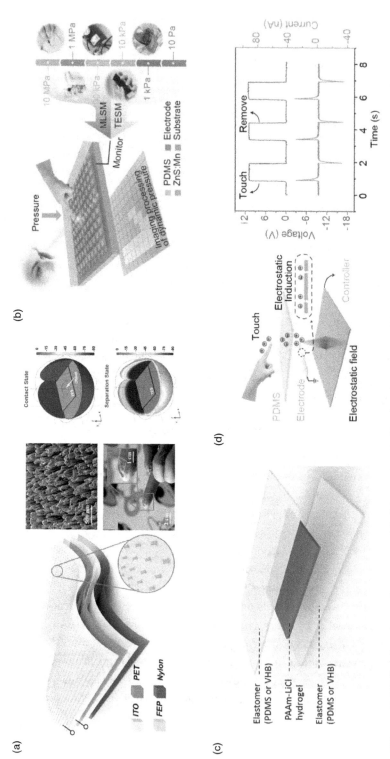

Figure 2.5 Triboelectric tactile sensors. (a) Schematic of triboelectric sensor with an electrification layer of aligned fluorinated ethylene propylene (FEP) nanowires, and 3D potential distribution of sensors at the contact state. Reproduced with permission from Ref. [27]. Copyright 2014 American Chemical Society. (b) A full dynamic-range triboelectric pressure sensor matrix based on the single-electrode

Caption for Figure 2.5 (cont.)

triboelectric sensor via using the coupling effect between contact electrification and electrostatic induction. Reproduced with permission from Ref. [69]. Copyright 2017 John Wiley and Sons. (c) A stretchable triboelectric nanogenerator for biomechanical energy harvesting and tactile sensing with hybrid elastomer and ionic hydrogel as the electrification layer and electrode, respectively. Reproduced with permission from Ref. [70]. Copyright 2017 American Association for the Advancement of Science. (d) A self-powered and stretchable triboelectric tactile sensor based on patterned Ag-nanofiber electrodes for detecting and spatially mapping trajectory profiles. Reproduced with permission from Ref. [71]. Copyright 2018 John Wiley and Sons.

Figure 2.5(a) gives the structural design of a self-powered sensor, SEM images of polymer nanowires by plasma dry etching, and 3D potential distribution at the contact state with contact and the separation state without contact, respectively [27]. Recently, Wang's group introduced a full dynamic-range triboelectric pressure sensor matrix, using the coupling effect between electrostatic induction and contact electrification [69]. The device is composed of two parts: a triboelectric sensor matrix and a mechanoluminescent sensor matrix, which is responsive to different detection ranges, illustrated in Figure 2.5(b). A 100 × 100 large-scale pressure sensor matrix at 100 dpi resolution possesses high sensitivity. Recently, Pu et al. introduced a soft skin-like triboelectric nanogenerator for biomechanical energy harvesting and tactile sensing [70], as illustrated in Figure 2.5(c). They hybridized elastomer and ionic hydrogel as the electrification layer and electrode, respectively, and achieved good stretchability of uniaxial strain (1160%) and good transparency of visible light transmittance (96.2%). Also, they improved the outputting alternative electricity with peak power density of 35 mW m^{-2}, providing the potential application in artificial electronic skin for pressure perception. Wang et al. employed the patterned Ag-nanofiber electrodes to make a self-powered and stretchable triboelectric tactile sensor for detection of trajectory profiles [71], as shown in Figure 2.5(d). And an 8 × 8 cross-type sensor array can work well under high strain because of electrostatic induction, and the resultant device can execute rapid tactile mapping with a response time of 70 ms. Their self-powered triboelectric sensor matrix is expected to have potential applications in tactile sensing.

2.5 Optical Sensor

By converting the applied pressure to optical information as the electrical output, optical pressure sensors are developed. For example, PDMS was utilized as a pressure-sensitive waveguide to introduce an optical pressure sensor with integrated organic light-emitting diodes (OLEDs) and organic photodiodes (OPDs) [73], which are shown in Figure 2.6(a). The device is capable of outputting a pressure-controlled light intensity. Another example of optical pressure sensor is shown in Figure 2.6(b). Wang et al. in Javey's group developed a user-interactive e-skin by using human-readable output to visualize the detected pressure [74]. Organic light-emitting diodes (OLEDs) were incorporated to output light with different intensities to quantify the amplitude of applied loads. Through the interconnection of the anode of an OLED and the drain of a TFT, the current flowing into the OLED could be effectively controlled by the resistance read from a pressure-sensitive rubber (PSR), and hence by the applied pressure. In this way, a visual pressure response was achieved, which

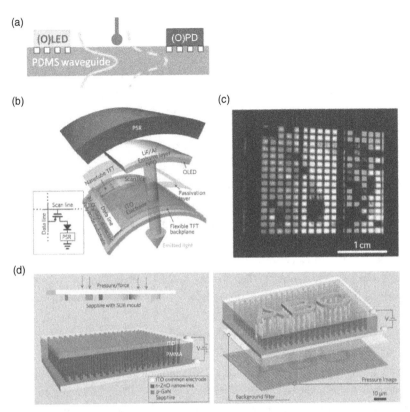

Figure 2.6 Optical tactile sensors and their working schematics. Reproduced with permission from Ref. [73], [74], [11]. Copyright 2012 John Wiley and Sons, 2013 Nature Publishing Group, and 2013 Nature Publishing Group.

is shown in Figure 2.6(c). Pan et al. also utilized the piezo-phototronic effect to make the optical pressure sensor, as shown in Figure 2.6(d) [11]. Patterned nanowires composed of ZnO and GaN were selected as piezoelectric semiconductors. When pressure was applied, nanowires were deformed to generate piezo-potentials, which could be used to control the performance of optoelectronic devices. In their design, pressure distribution could be mapped and a resolution of 2.7 μm was achieved. Besides, stretchable OLED was also studied [33], [73]. Elastic conductors and printable techniques were used to make large-area optical e-skins [32].

2.6 Other Sensing Mechanisms

2.6.1 Transistor

Thin-film transistors are also capable of functioning as a pressure sensor. When the pressure is applied, electrical properties (resistance or capacitance) at the gate or source-drain electrodes can be tuned. The change can be detected by measuring the currents from the transistor. An example is shown in Figure 2.7(a) [12]. Two separate functional layers were laminated as the pressure-dependent transistor. The first layer was source-drain electrodes and the semiconducting polymer. Polyisoindigobithiophene-siloxane (PiI2T-Si) was selected as the semiconductor because of its high mobility and other good electrical properties. The second layer was a micropyramidal PDMS layer on a PET/ITO substrate, which was regarded as the gate electrode. When the two layers were laminated, a pressure-sensitive capacitor was formed between them. Figure 2.7(b) shows the pressure-dependent performance of the transistor when constant source-drain and source-gate voltages were applied. This device has high potential to be developed as a pressure sensor. To access large-area fabrication of the stretchable e-skin, Someya et al. presented an engineering solution by processing organic transistor-based circuits to form a net-shaped structure [75]. By employing the unique structural engineering strategy, even materials that are stiff and not inherently stretchable could exhibit extensive stretchability for large-area networks. Figure 2.7(c) shows the network of transistor arrays and the circuit design for signal readout. However, the circuit-level stretchability through structural engineering requires sophisticated fabrication techniques, which limits its applications. Instead, the strategy of large-area fabrication of intrinsically stretchable materials attracted more interest. Wang et al. at Bao's group studied the scalable fabrication of transistor arrays utilizing intrinsically stretchable materials [13]. The schematic of one single transistor from the array is shown in Figure 2.7(d). Layer-by-layer deposition of stretchable components was used because of the resulting high yield and good uniformity of fabrication. For materials that are not incompatible with photolithography process, for example stretchable semiconductors, strategies like mask-protected etching and inkjet printing were developed for selective patterning.

2.6.2 Cracked Metal Film-Based Sensor

When a piece of thin metal film cracks, conductive paths will be disconnected. Resistance will increase correspondingly. The cracked metal film-based sensor can be regarded as a strain-gauge sensor, where the in-plane strains cause the change of electrical resistance on the cracked metal film. Based on this mechanism, a highly sensitive strain sensor on flexible substrate was designed by Kang

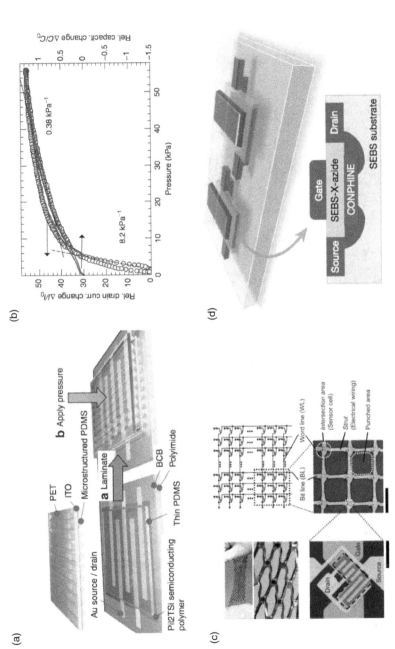

Figure 2.7 Transistor-based tactile sensors. (a), (b) Schematic of a polymer-based flexible transistor and its performance. Reproduced with permission from Ref. [12]. Copyright 2013 Nature Publishing Group. (c) Large-area network of e-skin based on transistor arrays and circuit design for structural stretchability. Reproduced with permission from Ref. [75]. Copyright 2005 National Academy of Sciences. (d) Stretchable transistor based on intrinsically stretchable materials. Reproduced with permission from Ref. [13]. Copyright 2018 Nature Publishing Group.

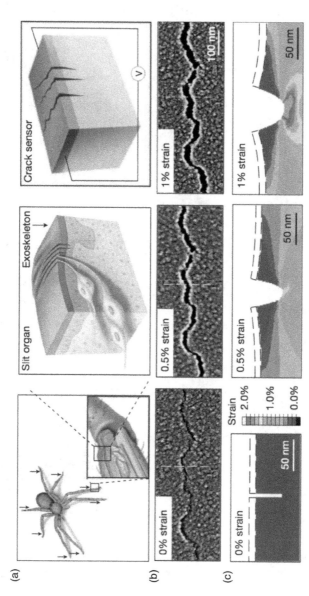

Figure 2.8 Spider-inspired crack-induced piezoresistive strain sensor. Reproduced with permission from Ref. [76]. Copyright 2014 Nature Publishing Group.

et al. to detect tensile forces. As shown in Figure 2.8(a), this work was inspired by the sensory system of a spider [76]. Slit organs, which are in an exoskeleton and a viscoelastic pad, in the vicinity of a leg joint of a spider, make spider legs sensitive to small force and vibration. Inspired by this structure, a 20 nm-thick platinum (Pt) layer was deposited on 10 μm-thick polyurethane acrylate (PUA). Metal cracks were designed by prestretching, through which the crack spacing and density could be controlled. Figure 2.8(b) shows the control of crack spacing at different strains. When the tensile force was applied, the cracked Pt films were disconnected and the resistance increased. A high gauge factor up to 2000 was acquired in a strain range of 0–2%. Low vibration with an amplitude of around 10 nm was also detected. Figure 2.8(c) shows FEA results, indicating the interfacial crack formation and separation at different strains.

This section discussed the various sensing mechanisms and recent progress of their resultant tactile sensors in the application of biomedical, robotics, and human–machine interface. A summary of diverse sensing mechanisms with related pros and cons is given in Table 2.1.

Table 2.1 Diverse sensing mechanisms of tactile sensors with relative pros and cons.

Type	Sensing mechanism	Advantages	Disadvantages
Capacitive	Detecting pressure via capacitance variation (i.e., the distance change between electrodes and/or deformation of dielectric materials)	1. Simple design of sensor structure 2. Low temperature sensitivity 3. Low power consumption	1. Complex design of circuit 2. Crosstalk issue for multiunit 3. Difficult for miniaturization and integration
Piezoresistive	Detecting pressure via electric resistance variation (i.e., the change of conductive path by quantum tunneling effect or percolation network)	1. Simple design of circuits 2. High sensitivity 3. Easy for miniaturization and integration	1. Nonlinearity of pressure response 2. Performance degradation over time 3. Low resistance to temperature and/or humidity

Table 2.1 (cont.)

Type	Sensing mechanism	Advantages	Disadvantages
Piezoelectric	Piezoelectric effect (i.e., piezoelectric potential generated in piezoelectric materials as a response to applied stress/ strain)	1. Production of electrical power or a signal by itself 2. Multitouch ability 3. Dynamic detection of pressure	1. Difficult to measure static pressure 2. High processing temperature (i.e., ~600°C for PZT) 3. High electrical poling voltage (i.e., 50–80 MVm^{-1} for PVDF film)
Triboelectric	Contact electrification (i.e., contact-mode by vertical charge polarization and sliding-mode by an in-plane charge polarization)	1. Easy fabrication 2. Large output voltage 3. Diversity of materials selection	1. Low output current 2. Lack of device durability and output stability 3. Lack of effect-ive packaging
Optical	Visualization of pressure change (i.e., integration of pressure-sensitive material to change electric current/ potential of optoelectronic devices)	1. Pressure visualization 2. Stability in harsh environment	1. Difficult for quantitative measure of pressure 2. Difficult for flexibility and miniaturization

3 Circuits and Systems for Large Sensor Arrays

As seen in the previous sections, there has been a tremendous volume of work to advance innovations in tactile sensing within the realm of materials science and engineering. However, in order to enable eventual applications in real-world use, these sensors may not be suitable unless we consider system-level constraints [37].

Compared to industrial robots that operate in structured environments, tactile sensing in humanoids must incorporate multiple sensing modalities and be able to cope with unstructured interactions in a real-world environment. Moreover, these tactile sensors must be tightly integrated to achieve high densities in certain areas and be able to cover a large area without sacrificing temporal resolution. These conflicting constraints pose several challenges in building a practical tactile sensing system [77], [78].

There are an estimated number of at least 45,000 mechanoreceptors over 1.5 m^2 area of representative human skin [79]. The distribution of these mechanoreceptors is uneven. In particular, the density of these touch receptors is highest in the human hand, where there are about 19,000 mechanoreceptors in the glabrous region (194 cm^2 area), and about 300 in the dorsum region (194 cm^2 area) [80], [81], [82]. Therefore, scalability of the architecture is extremely vital for successful implementation of large-area tactile sensing hardware.

In this section, we shall only talk about the issues pertinent to hardware. The organization of this section is as follows: Section 3.1 provides an overview of tactile sensing requirements. Section 3.2 describes several hardware architectures. Section 3.3 gives examples of sensor interface electronics that can be used.

3.1 Hardware Requirements

The main applications for tactile sensing are in prosthetics, industrial and medical/surgical robotics, and interactive devices (e.g., touch-enabled surfaces) [77]. Each of these applications would have a different set of design constraints and considerations. In this section, we shall highlight a few important system-level design considerations [37]–[39], [83].

Spatial Resolution and Density: The spatial resolution can refer to the distance between taxels or the smallest discernible feature size of the taxel array. The spatial density is the number of taxels per unit area. The resolution can be easily converted to density and vice versa. The spatial resolution is set by the required minimum detectable feature size for the application. Using the human finger as a blueprint for a robotic finger, the spatial resolution required would be 1–2 mm [78]. The two-point resolution (or discrimination), which is the minimum distance between two *separate* stimuli that is distinguishable, is directly related to the spatial density. For pattern discrimination, the movement of the finger (coupled with signal processing techniques) allows the effective spatial resolution to go down to 0.1 mm [78].

Sensitivity and Dynamic Range: The sensitivity is the minimum detectable force, while the dynamic range is the input range of force that produces a valid output. For commercial applications, such as touch screens, the sensitivity and dynamic range are determined by the required response when the user interacts with the screen in specific scenarios (e.g., different pressure levels can be used to vary the thickness of a drawn line in a sketching app). However, when the environment is unstructured, researchers have often relied on the human sense of touch as a guideline. Studies suggest that a range of 0.01 N to 10 N would suffice [37], [78].

Temporal Resolution: The temporal (or time) resolution of an array is the minimum *relative time difference* in electrical signal transmission that can be achieved *between* any two taxels. The time latency, on the other hand, refers to the time delay of the individual taxel's electrical signal transmission after the stimulus. To illustrate the difference between these two definitions, note that it is perfectly possible for a group of taxels with identical latencies of 10 ms to achieve less than 1 ms of time resolution. Studies suggest that relative spike timings in tactile afferents are in the order of 1 ms [84], [85]. Furthermore, the relative timing of simple binary data streams from an array of taxels can convey enough information [86], which can be used for classification. To achieve millisecond time resolution, taxel response times have to be designed to be less than 1 ms [37], [78].

Crosstalk: Crosstalk between taxels can originate from fringing fields [87], multiple conduction paths [86], [88]–[90], or mechanical coupling between taxels. Electrical crosstalk occurs when the signal transmitted on one channel produces an *undesired effect* on another channel. This type of crosstalk can be managed by circuit design techniques and proper layout of sensitive signal traces. Mechanical crosstalk refers to the mechanical coupling between taxels. This type of crosstalk can be minimized at the outset by material selection and suitable mechanical construction.

Sensor Characteristics and Nonidealities: Sensor nonidealities can refer to any unwanted phenomena such as noise, nonlinearity, hysteresis, or drift (over time) in the transfer function of the sensor. Nonidealities over the working range of the sensor can be compensated for, as long as the sensor characteristic is repeatable and stable. Noise can be managed by proper circuit design, and drift can be compensated for by using signal processing techniques or calibration [91].

Sensing Modes: A sensing mode is an *aspect* of a stimulus that can be measured. For example, light can be measured by its intensity or frequency. In tactile arrays, multiple modalities *may* be incorporated into the sensor array to measure normal force, shear force, temperature, and vibration – similar to

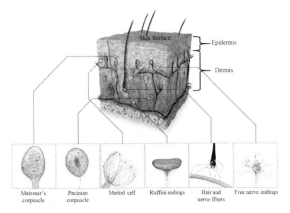

	Meissner Corpuscle	Pacinian Corpuscle	Merkel endings	Ruffini endings
Afferent type	FA-I (fast-adapting type I)	FA-II (fast-adapting type II)	SA-I (slowly-adapting type I)	SA-II (slowly-adapting type II)
Density (cm⁻²)	140	20	70	10
Frequency Sensitivity (Hz)	5-50 <ul><li>Insensitive to static force</li><li>Sensitive to dynamic changes</li></ul>	40-400 <ul><li>Insensitive to static force</li><li>Extremely sensitive to transients and high-frequency vibrations</li></ul>	<5 <ul><li>Sensitive to low-frequency and static forces</li></ul>	<5 <ul><li>Sensitive to low-frequency and static forces</li></ul>
Receptive field (mm)	3-5	>20	2-3	10-15

Figure 3.1 Diagram showing different tactile afferents in human skin [92] and a summary of its characteristics [93]. Image of skin taken from Ref. [92] Copyright 2018 by authors under CC-BY 4.0 license.

human skin. Nature has evolved a variety of specialized cells to do this. Some examples of these cells and their characteristics are shown in Figure 3.1.

Physical Constraints: Physical constraints arise from the form factor requirements for the application. Considerations include size, flexibility, and robustness (to the environment) of the sensor array. Moreover, as the number of taxels increases, wiring can easily become an issue. As a reference, a typical luxury car contains about 1.6 km of electrical copper wiring [94]. For large-area arrays, choosing a scalable architecture at the outset helps to ensure that wiring and hardware overheads are manageable in order to satisfy the required form factor.

Power: As there are many receptors in human skin, it is imperative for individual taxels to consume low power. Architectural decisions can also have a big impact on power consumption (e.g., a sleep mode). High energy efficiency of the tactile sensing system increases the options for selecting a power source and will extend operating time for battery-powered applications.

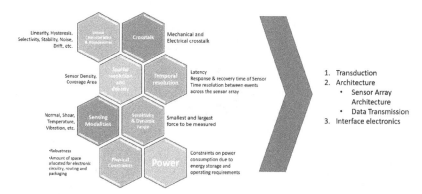

Figure 3.2 Summary of design considerations and challenges when designing hardware for tactile sensing systems.

The design considerations and its outcome are summarized in Figure 3.2. The decisions made will be reflected in the choice of sensors, hardware architecture [37], and interface electronics. An example, in the domain of robotics, is shown in Figure 3.3.

As shown in Figure 3.3, the design of a tactile sensing system can be broken down into the choice of transduction mechanism, architecture, and interface electronics. The touch sensor is a 5 × 5 POSFET-based array that must fit on the distal phalange of the robot finger. The electronics have been partitioned into two printed circuit boards (PCB) in order to satisfy the physical constraints. The touch sensor interfaces with the application-specific integrated circuit (ASIC) via an address bus and signal lines. The ASIC amplifies the analog signal from the sensor and conveys it to the analog-to-digital converter (ADC) located on the second PCB. The ADC on the second PCB digitizes this signal and transmits the result via a dedicated serial line to the main controller.

3.2 Sensor Array Architecture

3.2.1 Matrix Arrays

Most reported sensor arrays are laid out in a matrix (crossbar). Each taxel is read by asserting the row and reading out the columns (row-column decoding). Because of the decoding scheme, this type of sensor array usually comes in squares or rectangles. Matrix sensor arrays can be classified as passive [86], [88], [89], [96] or active [75], [97]–[101]. Both array types can achieve good spatial resolution, and readout circuity is relatively simple.

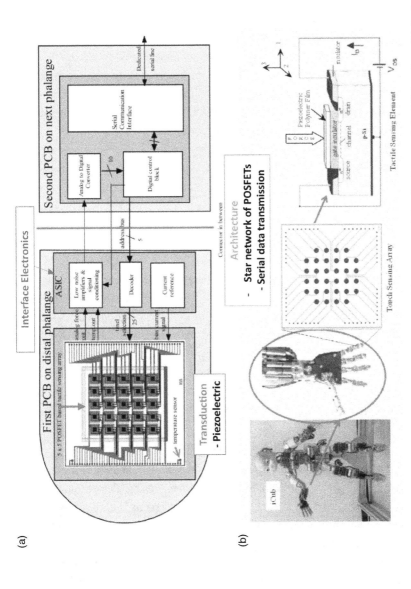

Figure 3.3 (a) Block-level description of the hardware system architecture used for the finger of the humanoid robot, iCub. (b) Left: Picture of the humanoid robot, iCub, showing the location of the sensor array on the phalange. Middle: Layout of the sensor array. Right: Vertical cross-section of the individual sensor element [37], [95]. Figure 3.3(a) from Ref. [37] and Figure 3.3(b) from Ref. [95].

Passive Matrix Arrays

Passive matrix arrays are typically used to implement piezoresistive [86], [88]–[90], [96] and capacitive [102]–[104] tactile sensing. To allow concurrent readout of each column and eliminate the inherent parasitic paths that are present in passive arrays, multiple op-amps are used to create a virtual ground to provide isolation [105], [106].

In Figure 3.4, some of the possible parasitic paths are shown for a tactile sensor matrix array. When the row is selected, neighboring row electrodes can be "grounded" (or a connected reference voltage) so that they do not propagate the unwanted signal paths (red and orange paths) across rows. Virtual grounds from inverting amplifiers allow parallel readout of columns by killing off unwanted paths that propagate across columns (blue).

Figure 3.5 shows some examples of passive matrix arrays from literature and one that is commercially available. Figure 3.5(a) has taxels with a spatial resolution of 2 mm on a flexible polyimide sheet, while Figure 3.5(b) uses piezoresistive fabric and achieves a spatial resolution of 4 mm. Figure 3.5(c) is a commercial piezoresistive array from TekScan [90]. These piezoresistive sensor arrays all use the same op-amp technique to reduce crosstalk between taxels.

In a capacitive array, the mutual capacitance is used to detect a touch event. As the finger comes close to the row-column taxel, the mutual capacitance C_m between the row and column changes (Figure 3.5e). The change in capacitance is measured to detect a touch event. For the capacitive array in Figure 3.5d, the column line is connected to a TIA, and adjacent columns are switched to ground to eliminate the parasitic paths. Industrial capacitive arrays are available from PPS [104], while commercial capacitive arrays are found in most mobile devices [103]. Arrays from TekScan and PPS have array sizes of more than 1,000 with sample rates that range from 10 Hz to 20 kHz.

Active Matrix Arrays

An active matrix array has taxels arranged in a crossbar (row-column) fashion like passive arrays. Each taxel consists of a *passive* sensing element and an *active* element that is tightly integrated. The active element is almost always a FET (field-effect transistor). The FET in the taxel eliminates the parasitic paths seen in passive arrays, without the need for an op-amp.

Figure 3.6 summarizes the different flavors of active matrix arrays seen in literature, according to the transduction mechanism of the taxel element.

In the two-terminal strain-gated piezotronic transistor [100] shown in Figure 3.6(a), the charge carrier transport is modulated by the piezoelectric polarization of charge across the vertical zinc oxide nanowire (NW). The induced

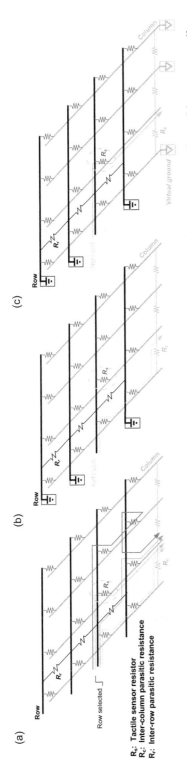

(a)

(b)

(c)

R_s: Tactile sensor resistor
R_c: Inter-column parasitic resistance
R_r: Inter-row parasitic resistance

Figure 3.4 (a) shows some signal propagation paths that contribute to crosstalk [88], and (b),(c) show how grounding mitigates crosstalk. Figure 3.4(a),(b), and (c) adapted from Ref. [88].

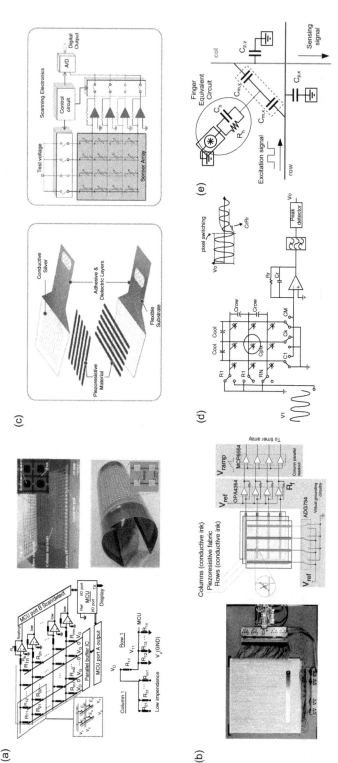

Figure 3.5 (a),(b),(c) Passive piezoresistive matrix arrays from [86], [89], [90] respectively. (d) Capacitive matrix array showing the electronics used for readout [102]. (e) The effect a finger has on the mutual capacitance between the row and column [103]. Figure 3.5(a) from Ref. [89] Copyright 2009 IEEE. Figure 3.5(b) from Ref. [86] Copyright 2017 by authors under CC-BY 4.0 license. Figure 3.5(c) from Ref. [90] Copyright Tekscan. Figure 3.5(d) from Ref. [102] Copyright 2002 IEEE. Figure 3.5(e) adapted from [103].

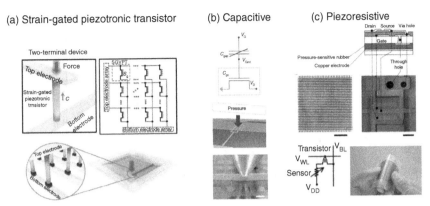

(a) Strain-gated piezotronic transistor   (b) Capacitive   (c) Piezoresistive

Figure 3.6 (a),(b),(c) Three different types of active matrix taxel elements with different transduction mechanisms. Figure 3.6(a) from Ref. [100] Copyright 2017 American Association for the Advancement of Science. Figure 3.6(b) from Ref. [99] Copyright 2015 by Nature Publishing Group. Figure 3.6(c) from Ref. [75] Copyright 2005 The National Academy of Sciences of the USA.

charge results in a potential across the NW that controls the Schottky barrier height and hence the transport characteristic.

In the capacitive active arrays [99] depicted in Figure 3.6(b), the capacitive sensing element is in series with the FET gate. In the suspended gate organic thin-film transistor (SGOTFT), pressure on the suspended gate modulates the capacitance of the dielectric and hence the drain-source current.

A similar type of taxel, with the sensing element at the FET gate, is the POSFET [107]. In this case, the sensing element at the gate is piezoelectric (i.e., not passive) and row-column addressing is not directly applicable. Each taxel must be connected directly to the readout module to form a star network in Figure 3.6.

In piezoresistive active arrays [75], [97], shown in Figure 3.6(c), a pressure-sensitive rubber (PSR) is placed in series with the FET. The resistance is modulated by pressure, and resistance values are read out via row-column addressing.

One of the most obvious drawbacks of active matrix arrays is the low carrier mobility of organic FETs (OFETs). This translates to high operating voltages (10 V~60 V) and low speeds (cycle times 1~10 ms) [75], [99]. However, NW-array FETs can exhibit carrier mobilities that are 10 times higher than OFETs and operating voltages of less than 5 V [97].

A summary of matrix arrays is given in Table 3.1.

3.2.2 Star Networks / Point-to-Point

An example of a capacitive implementation is shown in Figure 3.7 [108]. The transducer is a capacitive force sensor that has a middle plate, which is enclosed

Table 3.1 Summary of matrix arrays.

Reference	Substrate	Matrix	Transduction	Spatial resolution (mm)	Mobility ($cm^2\ V^{-1}\ s^{-1}$)	Sample rate (Hz)	No. of taxels
[96] Lee	Fabric	Passive	Piezoresistive	6	-	5.2 K	64 × 64
[88] Yang	Polyimide, copper	Passive	Piezoresistive	10	-	-	8 × 8
[89] Chang	Polyimide, copper	Passive	Piezoresistive	5	-	-	25 × 25
[8] Someya	Plastic film	Active	PSR, OFET	2.5	1.4	33*	16 × 16
[97] Takei	Polyimide	Active	PSR, NWFET	3.9	20	-	18 × 19
[98] Yeom	PET	Active	PSR, SWCNT	3.35	1.1	-	20 × 20
[99] Zang	PET	Active	Capacitive, SGOTFT	7.5	0.34	-	8 × 8
[100] Wu	Si wafer or PET	Active	Piezotronic, SGVPT	0.109	-	-	92 × 92

*Limited by transistor cycle time

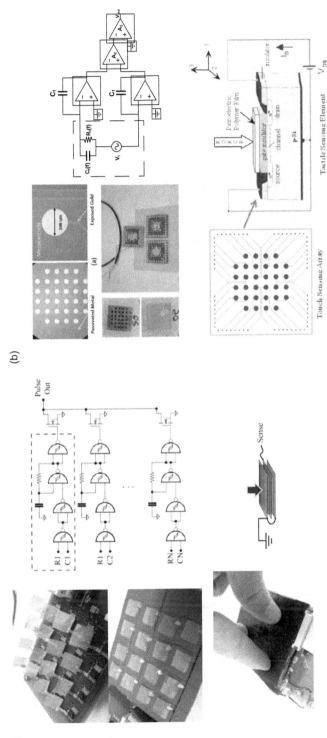

Figure 3.7 (a),(b) Examples of star networks from [95], [108] respectively. Figure 3.7(a) from Ref. [95] Copyright 2009 IEEE. Figure 3.7(b) from Ref. [108] Copyright 2010 IEEE.

on the top and bottom by two grounded metal plates. The grounded outer plates shield the center electrode from environmental noise and stray capacitances. Each taxel is addressed sequentially, with the value of the capacitance encoded into the frequency of the oscillator. The output of this oscillator is transmitted, via a single output, to the readout circuit. The single output simplifies wiring, but it also limits the speed of the array, as only 1 taxel can transmit at any time.

The implementation in Figure 3.7(b) has an array with 32 microelectrodes with piezoelectric taxels [95]. The taxels are connected in a star network with a spatial resolution of 1 mm. The immediate observation is the amount of wiring needed to route each individual taxel to the edge of the module so that it can be accessed by measurement equipment.

Both examples can be considered a star network, where the output (or addressing bits) of *each* individual taxel is connected to the readout circuitry (or decoder) by at least one metal trace. These taxels can be read serially by a single ADC to reduce power consumption or concurrently by multiple ADCs to reduce response time. Regardless, as the number of taxels increases, the number of wires increases linearly. Eventually, the amount of wiring limits the number of taxels that can be connected to a single module.

We will see that for scalable architectures, to reduce external routing, signal conditioning is performed locally with either discrete components or an application-specific IC (ASIC). This allows for a larger-scale implementation. To maintain good response times, each sensor module will usually be a star network with not more than 32 taxels.

3.2.3 Scalable Architectures

Today, electronic systems are being investigated to address many of the practical hardware issues related to the scalability of taxels over a large area. Some common themes among these systems are the use of send-on-delta (threshold) event-based signaling, asynchronous data transmission, and a shared communication bus.

Time-Multiplexed Bus

The approach taken by Cannata [109] for ROBOSKIN is to create smaller modular modules of 12 capacitive taxels each. On each module, each taxel is read out sequentially using a single capacitive sensing integrated circuit. The modules (up to 16) can be daisy-chained with a serial bus that terminates at a microcontroller. Each microcontroller then connects to a Controller Area Network (CAN) bus, which interfaces with a PC.

A more recent but similar modular approach is taken by CellulARSkin [110]. Each module is a hexagonally shaped "skin" cell that is equipped with a microcontroller and a variety of sensors (3 force, 1 accelerometer, 2 temperature, 1 proximity). Communication ports on the edge of the cell allow it to form an interconnected network with other cells via a fast 20 Mbits/s Universal Asynchronous Receiver Transmitter (UART) serial link (up to 66 cells). An interface box then connects the cell network to an Ethernet switch and subsequently to a PC. Overall, the 1 Gbits/s Ethernet connection can support 5742 cells. To reduce data load, an event-based signaling based on the *send-on-delta* concept is used.

Another event-based implementation by Tohoku University [111]–[113] uses a CSMA (carrier-sense multiple access) bus to connect multiple sensor chips. Each sensor chip contains integrated electronics for the sensor readout, logic, and data transmission. The logic also reduces the data load on the bus by transmitting only if the sensor reading exceeds a set threshold and by applying some form of adaptation [111]. In this case, the CSMA bus can achieve a higher temporal resolution (for the same bit rate) because it has a low overhead.

The systems mentioned above transmit sensor data in packets on a traditional type of communication bus (serial, UART, CSMA, Ethernet, etc.). Each packet would have to include the sensor data, sensor ID, and preamble bits. And since there is only one sensor node transmitting on the bus at any time, the temporal resolution of such a system would be constrained by the length of the data packet. The increase in the data collision rate as bus traffic (i.e., sensor activity) increases also results in a lower temporal resolution [111]. Recent work in [114] has improved the power, efficiency, and latency of the data bus by replacing Ethernet with an Asynchronous Serial Address-Event Representation protocol (AS-AER). The maximum transmission latency achieved was 340 ns (compared to 123 μ s in Ethernet).

Frequency-Hopping Multiple Access Bus

Instead of using a time-multiplexed communication bus, simultaneous transmission of sensor node data can be achieved by frequency hopping [115], [116]. In [115], each taxel consists of an FSR and a digitally controlled TFT LC oscillator (DCO). Variation in resistance modulates the *amplitude* of oscillation, and the digital control bits modulate the *frequency* of oscillation to implement frequency hopping. Frequency hopping allows more sensor outputs to transmit simultaneously for the same number of control bits vis-à-vis binary addressing [115]. This allows the sensor array to scale up tremendously without excessive wiring. However, as more sensing nodes are added to the network, the dynamic range requirement for the front-end electronics in the receiver increases.

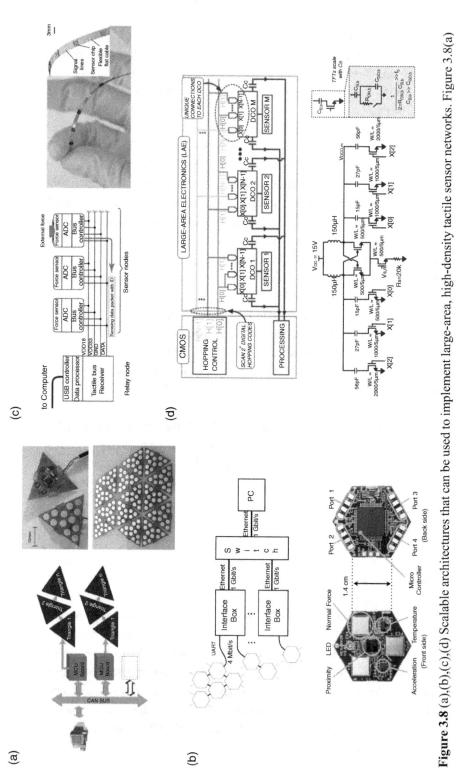

Figure 3.8 (a),(b),(c),(d) Scalable architectures that can be used to implement large-area, high-density tactile sensor networks. Figure 3.8(a) from Ref. [109] Copyright 2008 IEEE. Figure 3.8(b) from Ref. [110] Copyright 2015 IEEE. Figure 3.8(c) from Ref. [111] Copyright 2013 IEEE. Figure 3.8(d) from Ref. [115], [116] Copyright 2017–2018 IEEE.

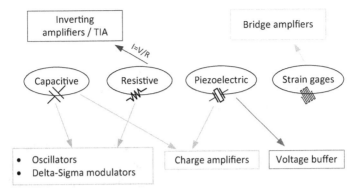

Figure 3.9 An overview of transducer types and the signal conditioning circuitry used.

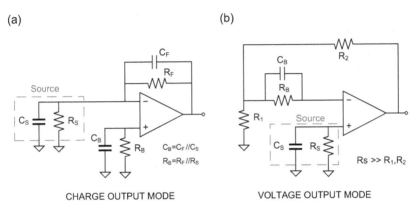

Figure 3.10 Charge output and voltage output mode amplifier configurations for transducers with large output impedances (adapted from [121]).

3.3 Interface Electronics for Signal Conditioning

This section provides a brief overview of the signal conditioning circuits used in the context of dense large-area sensor arrays. Ultimately, the choice of signal conditioning circuitry will depend on the transducer used for the sensor. A list of common transducer types and its associated signal conditioning circuitry is shown in Figure 3.9. The exact details and analysis of these circuits can be found in their respective references and also in textbooks [117], [118].

3.3.1 Charge, Voltage, and Transimpedance Amplifiers

Signal conditioning of transducers with high output impedances at DC (e.g., piezoelectric) can be accomplished in either charge output mode or voltage output mode [119], as shown in Figure 3.10.

In voltage output mode, the op-amp is in noninverting mode and acts as a voltage amplifier with high input impedance, as shown in Figure 3.10(b). If the source is a piezoelectric sensor in Figure 3.11(a), the charge generated during a contact event would result in a voltage across R_s that is amplified by the noninverting amplifier [119]. For the POSFET shown in Figure 3.11(b), the piezoelectric transducer is "natively" buffered by an NMOS voltage follower [120].

In charge output mode (charge amplifier), the amplifier is in the inverting mode and operates on the principle of charge conservation at the inverting input of the op-amp [122]. In Figure 3.11(c), the piezoelectric material is modeled as a Thevenin source, and the high-frequency gain, with respect to the input V_P, is C_p/C_f. The subsequent LPF is used to reduce the bandwidth and limit the output noise [122].

Schematically, the charge amplifier in Figure 3.10(a) is identical to the transimpedance amplifier and can be used for passive piezoresistive arrays as well (as seen in Figure 3.4). Transimpedance amplifiers can also be used to interface the current output from the active matrix arrays in Figure 3.6.

The charge amplifier configuration is also used for capacitive sensors. Since the capacitive sensor is not "self-generating" like the piezoelectric transducer, a carrier/excitation signal is required to measure the capacitance. In Figure 3.12 (a), the carrier amplitude is modulated by variations in amplifier gain due to the changing input capacitance, C_{pix}. In Figure 3.12(b), the switched-capacitor integrator measures the capacitance by first resetting C_2 and then integrating the charge Q_1 (on C_1) onto C_2 over a time period, T. The resulting output voltage, V_0, provides an averaged measurement of capacitance C_1.

The excitation signal can also be a sine wave signal, or it can be a switched DC signal as shown in the switched-capacitor implementation in Figure 3.12(c). An antiphase signal driving C_{DAC} effectively removes the fixed 20pF capacitance and ensures that the variable portion of the sensing capacitance is less than C_{ref}, thus maximizing the range of the signal conditioning circuitry. If only a single terminal is available, a four-phase fully differential switched-capacitor circuit like the one in Figure 3.12(d) can be used.

3.3.2 Oscillator-Based Circuits

Oscillator-based signal conditioning involves converting the variation in impedance (in response to the physical variable being measured) to a frequency. The use of oscillator-based signal conditioning (or, in general, a time-based approach) in tactile sensing draws parallels to how humans convey tactile signals from mechanoreceptors in the hands to the brain [85], [93].

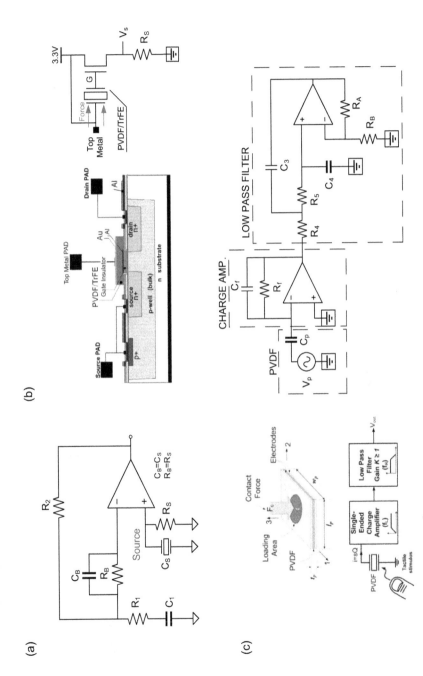

Figure 3.11 (a),(b),(c) Three different configurations of signal conditioning circuits for piezoelectric transducers: (a) voltage output mode, (b) a voltage follower, (c) charge output mode. Figure 3.11(b) from Ref. [120] Copyright 2011 IEEE. Figure 3.11(c) from Ref. [122] Copyright 2012 IEEE.

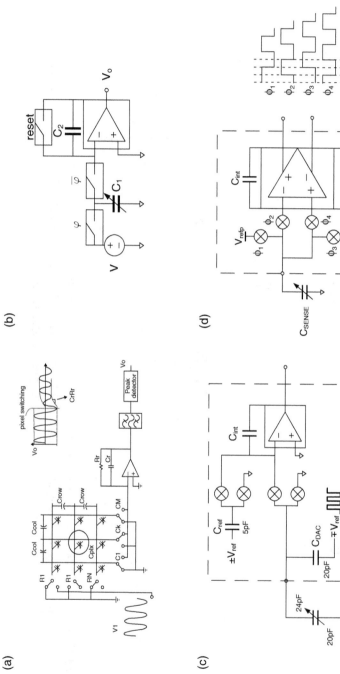

Figure 3.12 Signal conditioning circuits for capacitive sensors. (a) A matrix array used in conjunction with an amplifier in charge output mode. (b) A single capacitive taxel measured using a switched-capacitor integrator. (c) A capacitive taxel measured using a switched-capacitor integrator with a technique that extends dynamic range. (d) A capacitive taxel measured using a circuit that only requires a single-terminal connection. Figure 3.12(a) from Ref. [102] Copyright 2002 IEEE. Figure 3.12 (c),(d) from Ref. [123] Copyright 2005 IEEE.

Oscillators are usually used with capacitive and resistive transducers. Figure 3.13 shows various oscillator circuits used for sensor signal conditioning.

Figure 3.13(a) shows a ring oscillator that requires only a few transistors. The simple circuit can be implemented with organic FETs and integrated with the sensing element on the same flexible substrate [7]. Another architecture that is suitable is the relaxation oscillator, which uses a Schmitt trigger (hysteresis) to generate the required voltage threshold levels[108], [113] (shown in Figure 3.13 (a,b) as V_{SPH} and V_{SPL}).

Figure 3.13(d) shows another interesting approach that modulates both the frequency and duty cycle [124]. This technique uses fewer wires for data telemetry but requires more active components. For a highly dense and large-area sensor array, this is feasible only if one moves to a custom application-specific integrated circuit (ASIC) approach.

3.3.3 Delta-Sigma Converter/Modulators

$\Delta\Sigma$ (or $\Sigma\Delta$) data converters with moderate to high resolutions can be built using a coarse quantizer by exploiting signal processing techniques such as oversampling and feedback to shape the quantization noise [125]. This is attractive because the individual components will be easier to design. The 2nd-order modulator with a binary quantizer and an oversampling ratio (OSR) of 32 can achieve a 10-bit resolution [125]. For resolutions less than 10 bits, a 1st-order modulator like the ones in Figure 3.14 would suffice.

Figure 3.14 shows examples of capacitance-to-digital converters using discrete-time $\Delta\Sigma$ architectures. They are called discrete-time because the continuous-time input is sampled (using switches) before it enters the feedback loop. In Figure 3.14(a), the sensor capacitance is translated to an equivalent resistance, $R_{eq} = \frac{1}{f_{clk}C_{sense}}$, using the switched-capacitors techniques depicted in Figure 3.14(b) [126], [127]. This leads to an average input current , $I_{in} = \left(V_{dd} - V_{ref}\right)/R_{eq}$, that is integrated on the capacitor, C_0. If R_{eq} decreases (i.e., C_{sense} increases), I_{in} increases. This results in a larger feedback current through R_0 to maintain the ensure $V_+ = V_{ref}$ (on average). A larger feedback current means the FF (quantizer) stays high for a longer duration. The timer counts the duration that the FF stays high, thus providing a digital output of the capacitance value. Figure 3.14(c) is a 1st-order discrete-time $\Delta\Sigma$ modulator [123], except that V_{in} is known and the signal of interest is the input capacitance variation of C_{sense}. The circuit in Figure 3.14(c) can be modified easily for use with piezoresistive sensors by preceding it with a voltage follower. In fact, the switched-capacitor circuits in Figure 3.12 could be used for discrete-time $\Delta\Sigma$ modulators .

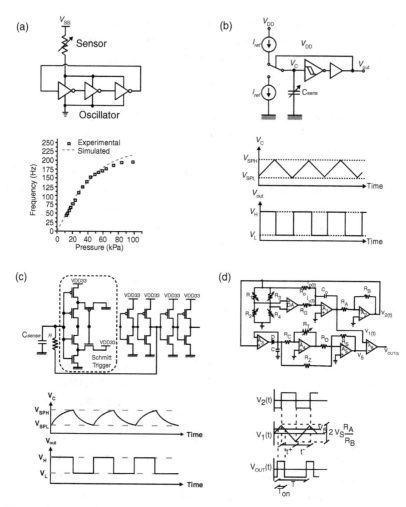

Figure 3.13 Oscillator circuits that were used to convert a physical quantity to frequency. (a) A voltage-controlled ring oscillator whose frequency is modulated by a piezoresistance. (b) A relaxation oscillator that uses a current source to charge and discharge a capacitive sensor to generate a frequency. (c) A relaxation oscillator for a capacitive taxel that uses a fixed resistance to charge and discharge the capacitive taxel. (d) A relaxation oscillator for a strain gage. The force from the strain gage modulates the output frequency, while a thermistor resistance modulates the duty cycle. Figure 3.13(a) from Ref. [54] Copyright 2015 American Association for the Advancement of Science. Figure 3.13(b) from Ref. [112] Copyright 2017 by authors under CC-BY 4.0 license. Figure 3.13(c) from Ref. [113] Copyright 2017 by authors under CC-BY 4.0 license. Figure 3.13(d) from Ref. [124] Copyright 1997 IEEE.

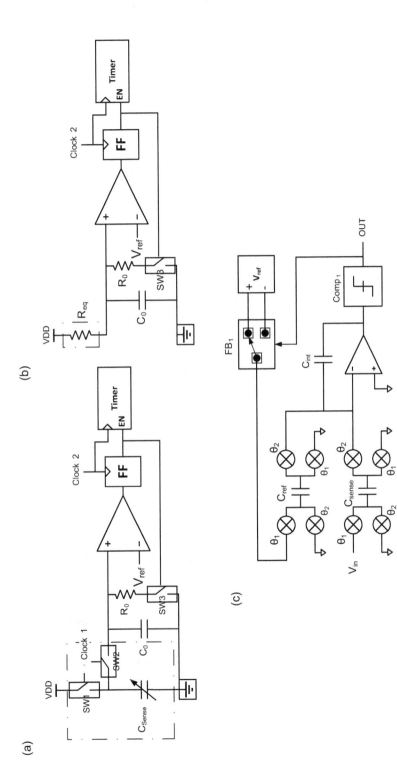

Figure 3.14 Discrete-time $\Delta\Sigma$ circuit topologies adapted from [123], [126] for capacitive sensors. (a) Schematic of a switched-capacitor implementation. (b) The equivalent circuit for (a) that shows how the sense capacitance acts as a resistor in the continuous time domain. (c) A 1st-order $\Delta\Sigma$ modulator. The modulator here uses an active-RC integrator compared to (a) and (b), which use a passive-RC integrator.

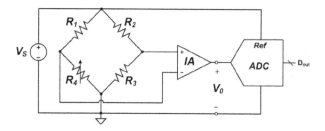

Figure 3.15 Example of a Wheatstone bridge (adapted from [117]) with piezoresistive sensor R_4 connected to a differential amplifier and ADC.

3.3.4 Bridge Amplifiers

Strain gauges produce only a small change in their resistance in response to the physical variable being measured. Bridge circuits allow measurement of these small resistance variations accurately by using a difference measurement (of two resistor dividers) instead of an absolute measurement. An example is shown in Figure 3.15. As the resistance variations are small, temperature variations can affect measurement if there is no proper compensation or calibration. To ensure that stray wire resistance does not affect measurements, the pair of sense terminals that connect the bridge output to the amplifier should not carry current. This means additional wiring is required to separate force and sensing terminals – Kelvin sensing [117]. Due to this complexity, strain gauges and bridge amplifiers are less commonly seen in dense large-area sensor arrays.

Fortunately, piezoresistive taxels have a large dynamic range, and most applications don't require a high resolution. Therefore, in most cases, simple resistor dividers are used. Regardless, there are still instances where strain gauges and bridge amplifiers have been used for touch applications that do not require high spatial resolutions [128]–[130]. For more details on the myriad bridge configurations and techniques, the reader can refer to [121], [117], [131].

4 Main Features and Trends of Tactile Sensors

The previous section provided an overview of the different components needed in electronic skins from the transducer to system-level integrations. In this section, we describe significant features and recent advancements of tactile sensors.

4.1 Sensitivity

Transducers are responsible for the perception of external mechanical stimuli, like strain, vibration, and so on, and result in a mechanical deformation. As a result, the mechanical deformation from the external tactile environment can be converted into electrical signals, which are processed by other parts of the e-skin system. Within the sensing process, sensitivity is a significant factor to quantify the performance of the sensor device, which is usually defined by the change of electrical signals (e.g., capacitance or resistance) divided by the change of pressure or strain. So far, there have been a number of effective approaches contributing to the progress of the sensitivity for tactile sensors. In this subsection, we review the recent advances of materials and structural designs for improving the sensitivity of tactile sensors in the field of electronics skins. The materials design will focus on the specific design of functional materials (e.g., their geometries and distribution) to enhance the conducting and sensing capabilities of the sensor, as well as the mechanical properties of polymeric materials. The structural design will focus on the improvement of structure for polymeric materials (e.g., their shape and sizes) for better deformability and hence for better sensitivity and conformability.

4.1.1 Materials Design

Polymeric materials and functional materials are essential components for pressure-sensitive tactile sensors, which require specific design for advanced sensing. The supporting layers of sensor structures are usually made of polymeric materials with a low Young's modulus for sensor flexibility. The sensitive layers are made of polymeric materials combined with functional materials. Polymers are designed to achieve the mechanical deformability of sensor structures when applied with pressure. Functional materials are designed for the electrical signal changes that occur when tactile sensors are deformed due to applied pressure. We will talk about the selection and designs of functional materials and the mechanical features of polymeric materials on sensor performances.

Functional materials like nanomaterials, conductive polymers, and metal thin films can be used in tactile sensors to convert the applied pressure to the change of electrical signals. In order to achieve superior sensitivity, functional materials are expected to be properly designed with specific surface morphologies. These morphologies could increase the contact area change when pressure is applied, improving the pressure response of tactile sensors. For example, a conductive layer coated on a fractured elastomer network improved the sensing performance [132]. Yao et al. designed a fractured Polyurethane (PU) network with conductive graphene nanosheets coated on it. PU sponge was first dip-coated in graphene oxide solutions to form the

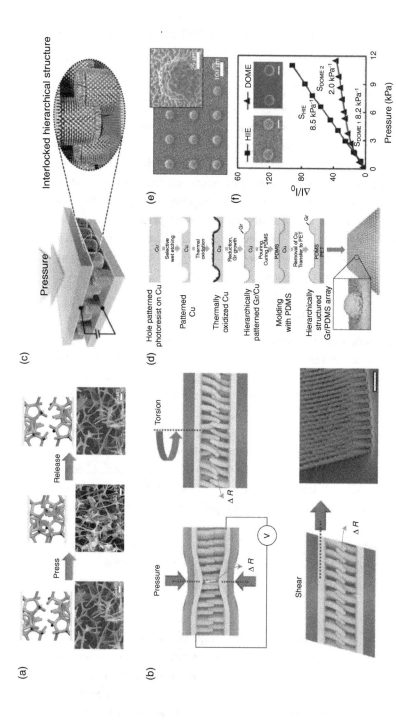

Figure 4.1 Functional materials design for enhanced sensitivity. (a) Fractured network of RGO-PU sponge. Reproduced with permission from Ref. [132]. Copyright 2013 John Wiley and Sons. (b), (c) Metal coating on interlocked elastomer. Reproduced with permission from Ref. [133], [1]. Copyright 2012 Nature Publishing Group and 2015 John Wiley and Sons. (d)–(f) Design of hierarchical structural Gr/PDMS array for electronic skin and its sensing performance. Reproduced with permission from Ref. [134]. Copyright 2016 John Wiley and Sons.

conductive layer on its surface. A reduced graphene oxide layer was formed after drying and applying the hydrogen iodide (HI) solution treatment on coated graphene nanosheets. After the hydrothermal treatment, RGO-PU sponge was hard compressed to generate the fractured network. Such a fractured network enables the sensing of applied pressure due to the change of contact area between conductive layers. Figure 4.1(a) illustrates the schematic and SEM images of the fractured RGO-PU sponge when it is pressed and released. The increase of contact area induced by applied loads leads to the reduction of resistance. As the fractured network has an enhanced deformability compared to the entity structure, such design of functional materials can significantly improve the sensor pressure response. For piezoresistive sensors, the differential of $(R_0 - R_p)/R_0$ versus pressure is defined as the sensitivity, where R_p and R_0 refer to the resistance of the sensor with and without loads, respectively. Fractured RGO-PU sponge sensor can achieve a sensitivity of 0.26 kPa^{-1}, showing an enhancement of two orders of magnitude compared to the sponge sensor without fractured structure, in the pressure range of 0–2 kPa. Pang et al. and Ha et al. also coated the functional materials on three-dimensional (3D) arrays. They developed the metal coating on interlocked elastomers. Sensor structures are shown in Figure 4.1(b) and Figure 4.1(c), [1], [133]. In both studies, micropillars were casted from the hole-patterned Si mold. As shown in Figure 4.1(c), ZnO nanowires were grown on the surface of PDMS pillars. They were fabricated using hydrothermal methods. Metal films were then coated on the surface and interlocked to function as pressure sensors due to the change of the contact area between different pillars.

Another example for the design of functional materials is the bioinspired hierarchical elastomer dome and graphene monolayer work carried out by Bae et al. [134]. They designed the surface morphology and distribution of functional graphene materials by patterning the substrate. This work used wet etching to create a dome-shaped structure on the copper sheet. The copper sheet was then thermally oxidized, reduced, and annealed to form hierarchical nanostructures on its surface. After the growth of the graphene monolayer, PDMS was molded and lifted off to get the hierarchically structured Gr/PDMS array. The fabrication process and structure geometry are shown in Figure 4.1(d) and Figure 4.1(e), respectively. Because of the small protuberances on the hierarchical Gr/PDMS structure, the number of contacted protuberances and their contact areas with electrode increased with pressure, compared to the smooth Gr/PDMS surface. As a result, the total contact area of the structure increases linearly with applied pressure, instead of the increase with the power of 0.67 for smooth structure. This can lead to the improvement in sensitivity over a wide pressure range. Figure 4.1(f) shows the pressure response of a hierarchically structured Gr/PDMS sensor. Enhanced sensitivity (8.5 kPa^{-1}) with a wide linear range (0–12 kPa) is exhibited.

The conductivity and sensing capabilities were also improved by designing the functional composites with silver flakes and multiwalled carbon nanotubes with self-assembled silver nanoparticles as sensor materials [135]. Nickel foam template was selected to develop the graphene porous network for highly sensitive pressure and strain sensors [136]. Graphene film was also designed to form a honeycomb-like layer with a bubble-decorated structure to achieve high sensitivity [137]. In order to only detect compressive loads for the flexible tactile sensor, Someya's group worked on a sensitive pressure sensor, which is, however, bending-insensitive. The sensor responds only to compressive pressure. Graphene and CNT were designed into nanofibers with nanoporous structures for this sensor [138].

Besides the design of functional materials, the mechanical properties of polymeric elastomers (e.g., Young's modulus used for the sensor) also affect the sensitivity and sensor performance. The Young's modulus influences the mechanical deformability of the sensor structure. This is important for elastomers that have tunable Young's modulus like PDMS. Lower Young's modulus leads to a higher deformation ability of PDMS. By tuning its Young's modulus, the sensitivity of tactile sensors can also be tuned. Many factors affect the modulus of PDMS. For example, the curing temperature of PDMS will affect its Young's modulus. Higher curing temperature results in a stiffer material, and a linear relationship between the Young's modulus and the curing temperature was observed [139]. Different ratios of PDMS base to curing agent can lead to different Young's moduli of PDMS. PDMS (Sylgard 184) shows highest Young's modulus when the ratio is 9:1 [140]. This property is important for the tuning of PDMS stiffness for different sensitivity requirements. Also, PDMS with different mechanical properties could be mixed to achieve the optimized result [141]. After the PDMS is mixed and cured in specific conditions, the strain rate of loads on the material will also affect its mechanical results. Higher strain rate usually increases the measured Young's modulus of PDMS [140], [142].

4.1.2 Structural Design

Polymeric microstructures have been extensively utilized in designs for improving the sensitivity of flexible sensors. As the pressure is equal to the applied force divided by the contact area, the microstructured elastomers can lead to a concentrated stress distribution at smaller-area regions. This generates the highly deformability of the sensor and hence high sensitivity, compared with sensors consisting of materials without microstructure. The structural properties and parameters of these elastomers are essential for the sensitivity and other performances by affecting the deformability of the sensor. When functional materials are coated on elastomers, different structures of elastomers also lead to different sensing behaviors.

In order to create microstructures, lotus leaves, mimosa, sandpaper, and other templates from nature and lives have been used as molds [146]–[148]. By taking advantage of the shape, size, and height difference of microstructures generated on the surface, improved sensitivity of the sensor can be exhibited compared with the nonpatterned one. However, microstructures made from these molds usually show irregular patterns on the whole sensor film. This will make the sensor less controllable in its performance. The random distribution of micro-structures, and their shapes and sizes, also induces huge performance differences for different sensors. Silk was also selected as a mold to create microstructures on PDMS to make a sensitive pressure sensor [149]. Compared with previous molds, silk can show a regular interweaved pattern on its surface to form uniform microstructures when used as a mold. Also, the size, density, and geometric aspect of the microstructure can be tuned by choosing silk with different knitting types. High sensitivity (1.8 kPa^{-1}), low detectable pressure limit (0.6 Pa), fast response time (<10 ms), and high stability (>67,500 cycles) could be achieved when microstructures created from silk mold were used in the sensor. However, custom microstructure designs still cannot be made from these predetermined patterns. For example, the spacing of different microstructures is not adjustable. Also, how these interwoven patterns affect the sensor pressure response is still not clear.

A more controllable way to create microstructures is to use Si mold, which has hole patterns on the surface. Custom microstructures can be patterned on a Si wafer. Such Si mold is fabricated by using lithography, dry etching, and wet etching. Size, geometry, and spacing of microstructures can be specifically designed and controlled. Park et al. fabricated different microstructures (pyramid, dome, and pillar) by making corresponding Si molds and studied their selective sensitivity to multidirectional forces (normal, shear, tensile, and bending) [143]. Figure 4.2(a) illustrates micropillars, micropyramids, and microdomes fabricated by MWNT/PDMS composite. Uniform pattern is exhibited with controllable geometric conditions from custom design. Two microstructured films were interlocked and assembled to form a piezoresistive sensor. This is to facilitate the exhibition of the selective sensitivity effect of different microstructures on multidirectional forces. Figure 4.2(b) shows the FEA modeling results of a localized stress distribution comparison among different

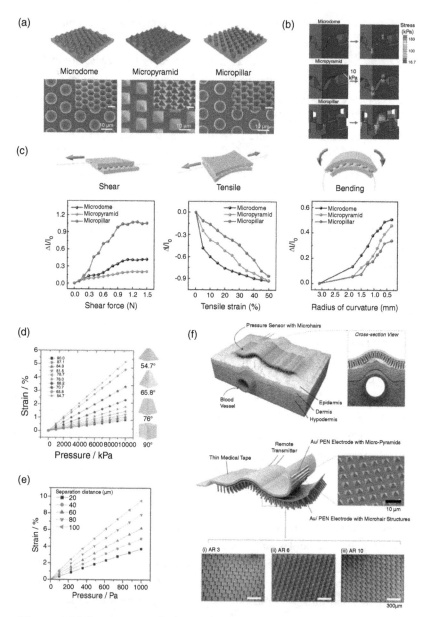

Figure 4.2 Microstructure design and geometrical parameters for enhanced sensitivity. (a), (b), (c) Effect of different microstructures on multidirectional forces sensing. Reproduced with permission from Ref. [143]. Copyright 2018 Nature Publishing Group. (d), (e) Study of geometrical conditions of micropyramid. Reproduced with permission from Ref. [144]. Copyright 2014 John Wiley and Sons. (f) Skin-conformal microhairy sensor with combined microstructures. Reproduced with permission from Ref. [145]. Copyright 2015 John Wiley and Sons.

sandwiched microstructures. It indicates the enhanced sensitivity effect of microstructures. Figure 4.2(c) shows the current change of sensors consisting of different microstructures when shear force, tensile strain, and bending force were applied. Different force and strain responses indicate the selective sensitivity of these microstructures on multidirectional loads. As a result, microdome structures present the highest sensitivity to normal, tensile, and bending force. Micropillar structures show the highest sensitivity to shear force.

Besides the study of different microstructures and their selective sensitivity, Tee et al. also explored the effect of geometrical parameters on sensitivity, as shown in Figure 4.2(d) and Figure 4.2(e) [144]. Their microstructures were fabricated using a soft PDMS mold made from soft lithography, which requires low fabrication cost and is good for large-area processes. This research work focused on the geometry effect of micropyramids on sensitivity. First, the sidewall angle of the microstructure was studied. When angle was decreased from 90 degrees to 54.7 degrees, the microstructure was changed from cube to pyramid. Micropyramidal structures indicated the lowest effective modulus and best sensitivity. In addition, the spacing of the two pyramids also affected sensitivity. This is from the change of the pyramids' number to tolerate applied load within a unit area. Pyramid patterns with different spacing were studied. Larger spacing leads to higher sensitivity. Based on this, sensing performance could be easily adjusted by designing different patterns. A mixed-sized pyramid pattern was also presented for tunable sensitivity. In order to apply these microstructures to sensor design, Bao's group incorporated both micropyramid and microhair structures to develop a skin-conformal pressure sensor [145]. Figure 4.2(f) illustrates their design. Consecutive layers of polyethylene naphthalene (PEN), Cr/Au electrode, and polyvinyl alcohol (PVA) were laminated with micropyramid array and microhair array, respectively, to form two functional layers. The two layers were assembled and sealed using thin medical tape as a bandage-like pressure sensor. Micropillar structure was also used to achieve a highly sensitive pressure sensor with a low detection limit [150].

To further improve the sensing performance of microstructures, hollow-shaped geometry shows enhanced sensitivity. For example, a spherical shell was demonstrated to exhibit considerable deformation ability [153]. It indicated promising use to be applied in microstructures for sensor application. Based on this, hollow-sphere structures of polypyrrole (PPy) were developed by Bao's group for a highly sensitive piezoresistive pressure sensor [151]. The fabrication was achieved through a multiphase synthesis technique. Figure 4.3(a) shows the schematic of hollow-structured PPy, indicating enhanced elasticity and good deformation ability. Figure 4.3(b) shows the schematic and Transmission Electron Microscopy (TEM) image of interconnected PPy with hollow sphere structures. PPy, which is stiff and brittle

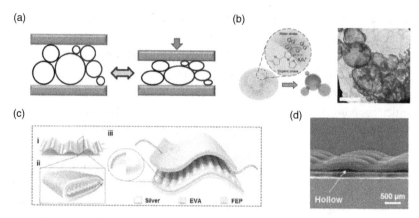

Figure 4.3 Development of hollow-patterned structure for sensitivity improvement. Reproduced with permission from Ref. [151], [152]. Copyright 2014 Nature Publishing Group and 2018 American Chemical Society.

due to the rigid conjugated-ring backbone, shows low effective elastic modulus under spherical shell structure. This makes it capable of withstanding a considerable force and showing good deformability and sensitivity. Pressure sensors containing this hollow-sphere PPy can show ultrahigh sensitivities of 7.7–41.9 kPa^{-1} within 100 Pa of pressure range. By further patterning the surface structure of PPy film, sensitivity was improved to 7.7–41.9 kPa^{-1} within a low-pressure regime of 30 Pa.

As illustrated in Figure 4.3(c), Chen et al. also developed hollow structures on the surface of microarrays [152]. Figure 4.3(d) shows the SEM image of the hollow shape on the surface of ethylene-vinyl acetate (EVA) material. This hollow microstructure-based pressure sensor was integrated into a noncontact heartbeat and respiration monitoring system to show high sensitivity (18.98 V/kPa) and wide sensing range (40 kPa). Other microporous structures were also studied to improve the sensing ability [154], [155]. They are promising in achieving sensitive detection of very low pressure. However, irregular distribution, geometric shape, and size of hollow patterns developed for now are still the limitations for the reproducibility of these sensors. A more controllable design strategy still needs to be explored.

4.2 Stretchability

The ability to stretch devices and electronics has opened new classes of applications in large-area electronics, because it provides opportunities to develop smart wearable devices, epidermal (on-skin), bioimplanted electronics, and robotics [4], [5], [156], [73]. In general, the electronics needs to not only bend, but also undergo deformations such as twisting, wrinkling, creasing, or folding. Thus, stretchability enables an intimately conformable surface of complex-shaped static/dynamic objects to possess high durability and robustness [157], [158], [159], [160].

In general, there are several approaches to make electronic devices stretchable. The first method uses the rigid device islands and stretchable interconnects based on the strain engineering. Herein, inorganic metals and semiconductors are regarded as optimal materials to maintain high conductivity subject to external strain [4], [161], [162]. In this method, all the applied strain is sustained by interconnects, whereas the isolated rigid islands experience no significant strain. On the other hand, microstructuring designs require electronic materials to present good flexibility, but the conventional bulk metal or semiconductor materials are usually rigid and brittle. Thus, nanometer-thick films are engineered to reduce structural stiffness [163], [50]. Alternatively, intrinsically conductive soft materials such as polymers can also be investigated as stretchable electronic materials candidates.

Highly sensitive pressure sensors have promising applications in electronics skins, wearable electronics, and soft robotics. Previous research mainly focused on flexibility; yet the stretchability of material and structure of tactile sensors is generally limited, which becomes difficult when trying to satisfy the new request of small size and conformability for updated soft electronic devices.

To date, there have been some effective strategies of materials design to improve the stretchability of the tactile sensor in addition to having flexible and sensitive features. Firstly, in the past decade, the success of stretchable electronics relied on mechanical designs in electronic materials and structures to allow them to experience various in-plane deformations that often result in out-of-plane or lateral buckling and wrinkle of thin-film components [157], [168]. Thus, harnessing buckle/wrinkle mechanisms of thin-film materials to create stretchable tactile sensors is still one robust tool today. For instance, Wei et al. fabricated piezoresistive fibers with multiscale wrinkling microstructures, and the resultant piezoresistive core-shell fibers present good stretchability (400%) and electrical conductivity (10^{-4}~10^{-5} Ω cm) [164]. The multiscale wrinkled microstructures were accomplished on the surfaces of fibers to increase the micro/nanostructures by writing silver nanowires, where AgNWs/waterborne polyurethane (WPU) composites layers were coated on prestrained PU fibers, illustrated in Figure 4.4(a). The significant strength of this approach can overcome the viscoelastic delay of polymer composites and thus reduce the response and relaxation times (35 and 15 ms). The stretchable piezoresistive conductive fibers, thus, can be a good candidate for future electronic skins and robotic systems.

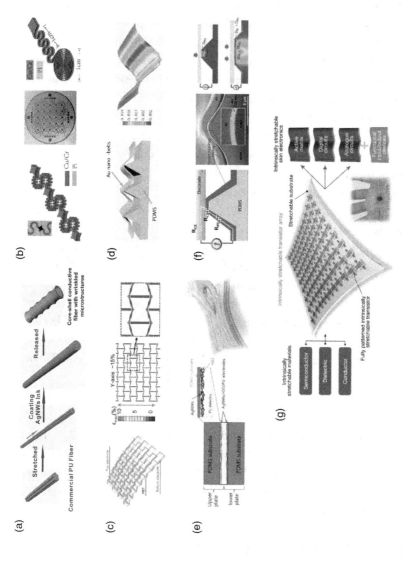

Figure 4.4 Designing strategies of stretchable tactile sensors. (a) Stretchable wrinkle/buckled microstructures: illustration of multiscale wrinkled microstructures based on stretchable core-shell piezoresistive fibers. Reproduced with permission from Ref. [164]. Copyright 2016

Caption for Figure 4.4 (cont.)

John Wiley and Sons. (b) Stretchable serpentine/fractal electrodes: second-order serpentine design based on stretchable capacitive pressure sensors by the interconnect wires. Reproduced with permission from Ref. [165]. Copyright 2017 Royal Society of Chemistry. (c) Stretchable origami/kirigami electrodes: a stretchable mutual capacitive pressure sensor array based on mesh-type auxetic structure and graphene electrodes. Reproduced with permission from Ref. [49]. Copyright 2017 American Chemical Society. (d) Stretchable sinusoidal nanobelt electrodes: design of out-of-plane tripod PDMS microstructure-based tactile sensor based on wavy stretchable gold nanobelt electrodes. Reproduced with permission from Ref. [166]. Copyright 2015 John Wiley and Sons. (e) Stretchable conductive nanomaterial film: stretchable and transparent capacitive touch sensors based on patterned Silver Nanowires/ Reduced Graphene Oxide (AgNWs/rGO) electrodes. Reproduced with permission from Ref. [167]. Copyright 2017 American Chemical Society. (f) Intrinsically stretchable conductive film: schematic of a stretchable piezoresistive sensor using a polymer-based stretchable electrode incorporating conductive polymer PEDOT:PSS and PUD elastomer blend. Reproduced with permission from Ref. [61]. Copyright 2014 John Wiley and Sons. (g) Large-area intrinsically stretchable sensor array: design of an intrinsically stretchable transistor array as the core platform of skin electronics. Reproduced with permission from Ref. [13]. Copyright 2018 Nature Publishing Group.

Apart from dispersing conductive nanomaterials (i.e., metal nanowires, carbon nanotubes, or graphene) into elastomeric matrices, an alternative strategy of improving stretchability is to create 2D filamentary meshing, which has the capability of cointegrating material process platforms with operation conditions and good spatial resolution. The main advantage of this approach is to combine the large-range force for active measurement, with the entire stretchability dependent on a soft matrix. Serpentine/fractal-type structures are regarded as self-similar, defined as tiny similar components generating approximated geometries [159], [169]. As a result, this self-similar feature endows separate active components with extensive elastic strain along the selected dimension and thus enlarges the stretch strain of large-area pressure sensor arrays. Instead of using out-of-plane connects or buckle patterns, in-plane type patterns are also employed to generate high stretch in rigid materials. In general, the serpentine wire can deform out of plane to increase the stretchability and meanwhile decrease the deformation. For example, Wang et al. utilized the second-order serpentine design to fabricate the interconnect wires for gaining high stretchability and deformability [165], as shown in Figure 4.4(b). Accordingly, two kinds of stretchable capacitive sensors were created, including planar film sensors and circular film sensors. And to achieve better conductivity and connecting when applying tensile strain, the connecting wires and electrodes are made with a sandwich capacitor structure (Cu/Cr/PI) on PDMS substrate. Then, a 3 × 3 PPS sensor array was stuck onto a robotic finger to differentiate surfaces and stiffness with the assistance of analyzing output signals, and the results show the promising application of flexible and stretchable capacitive devices in surgical robots.

Furthermore, origami (paper-folding) and kirigami (paper-cutting) based auxetic materials are another effective strategy to add mechanical stretchability to tactile sensors for further extending their range of wearable electronics applications [170], [171], [172]. The perforated auxetic structures can transform the external stretch strain into local bending motions of constituent beam elements, yielding an improved stretchability in pressure sensors. The unique advantage of the perforated auxetic structure is its ability to simultaneously generate stretching in both the X and Y directions under applied strain. For example, human palms expand biaxially when stretching strain is applied. Therefore, tactile sensors designed by auxetic structure can significantly improve the stability of sensor arrays, and thus could be attached to various human body parts. Recently, Kang et al. reported a wearable and stretchable mutual capacitive pressure sensor array based on mesh-type auxetic structure and graphene electrodes including multipoint-touch sensing and 3D sensing in high deformation state [49], as depicted in Figure 4.4(c). The resulting graphene-based 3D tactile sensor exhibits simultaneous 8% stretching in the X-axis and 15% tensile strain in the Y-axis. Moreover, the use of graphene electrodes enhances stretchability and transparency and minimizes mechanical failure during operation. As a result, the thinness and mechanical stretchability of the overall sensor can be used for a variety of remote assistive communication devices by generating conformal contact with highly deformable human body parts.

To overcome unstable conductivity and limited stretchability of dispersing conductive nanomaterials into elastomeric matrices, zigzag/sinusoidal-shaped metal films/belts located on the surface of the elastomer can be designed to achieve stretchable conductors. For example, Qi et al. employed wavy, stretchable gold nanobelts as electrodes to fabricate an out-of-plane tripod PDMS microstructure-based tactile sensor [166], as shown in Figure 4.4(d). The resultant sinusoidal-shaped electrodes can bear 130% stretching and repeated stretch/relaxation over 10,000 cycles with little increment of the belt resistance. Also, the gold nanobelts are still adhesive to the microstructure film when stretching is applied; and, after the stretch is released, a sinusoidal texture presents, owing to the buckling in the elongation direction. So, the resulting flexible and stretchable electrode integrated in tripod elastomeric substrate enables a comprehensive method to design stretchable pressure sensors for further stretchable electronics and biointegrated electronics.

Additionally, Choi et al. embedded stretchable silver nanowires on a PMDS substrate and subsequently made a transparent and stretch-unresponsive stretchable capacitive sensor [167]. Figure 4.4(e) demonstrates the schematic of a stretchable capacitive touch sensor. These devices break through the limit of capacitive touch sensors, in which the capacitance changes result from geometry changes of hyperelastic dielectrics, and thus it is difficult to achieve the stretchable electrodes that keep stable under large deformations. The achieved stretch-unresponsive touch sensing capability of this AgNWs/rGO electrode-based stretchable capacitive device might be applied in the field of human–machine interfaces.

With the increasing complexity and multifunction of flexible electronics, intrinsically stretchable, conductive polymer films could be good candidates for stiff electrode material with a brittle nature and high processing and thus hopefully lead to next-generation human–machine interactive electronics and robotics [173], [25]. The previous sheet resistance of conductive polymer PEDOT:PSS (10^5 Ω/sq) is so high that it cannot be used as electrode material, but Bao's group significantly decreased the sheet resistance of PEDOT:PSS to 240 Ω/sq via incorporating additives of fluorosurfactant Zonyl and dimethyl sulfoxide at 97% transparency [174]. The increase of phase separation enhances electrical conductivity between the conductive PEDOT and the insulating PSS chains, resulting in increased conducting paths of PEDOT. On the other hand, harnessing highly sensitive micropyramid PDMS arrays can well combine resistivity with the stretchability of PEDOT:PSS as a conductive electrode. For example, Choong et al. introduced a polymer-based stretchable electrode including a conductive polymer PEDOT:PSS and an aqueous polyurethane dispersion (PUD) elastomer blend [61]. As seen in Figure 4.4(f), herein a PDMS substrate with micropyramids can be fabricated with a thin PEDOT: PSS/PUD blend, which serves as a piezoresistive electrode. The resulting tactile sensor has better pressure sensitivity with 10.3 kPa^{-1} at 40% tensile strain, and its potential usability in an intrinsically stretchable tactile sensor as an E-skin sensor was tested to be sensitive enough to measure the small force subjected to stretching.

Finally, stretchable electronics generally consists of two kinds of components: one is input/output devices for human interaction (i.e., a pressure sensor element for input and an electrical signal change for output); the other one is electronic circuits for information processing. There have been numerous efforts devoted to signal input and output via increasingly sensitive and stretchable sensors so far, but the studies on stretchable circuits are still few. The recent work by Wang et al. introduced a fabrication process with high yield and uniformity from various intrinsically stretchable conductive polymers, where a stretchable transistor array with device density of 347 transistors per square centimeter is realized without current-voltage hysteresis after 1,000 cycles [13]. Figure 4.4(g) illustrates the 3D morphology of an intrinsically stretchable transistor array for electronic skin. Thus, this comprehensive fabrication platform for designing intrinsically stretchable transistor arrays has the potential to move materials development systematically to desired electronic applications. The success of integrated intrinsically stretchable sensors and circuit arrays could lead the advancement of electronic materials to higher complexity at both circuit and device levels for the applications of skin-like electronics and even beyond-skin softness and deformability.

In this subsection, the various strategies and material systems of enhancing stretchability in tactile sensors were discussed. A summary of methods, materials, and structures of stretchability is given in Table 4.1.

4.3 Self-Healing

To mimic human skin and muscles that are capable of autonomously repairing themselves, several self-healing polymers have been developed in recent years

Table 4.1 Strategy of stretchability enhancement with various materials and structures.

Strategy	Structure	Material	Stretchability
Strain engineering	Wrinkled/buckled microstructure	1. Silicon nanoribbon on PDMS substrate 2. Metal coating polymer bilayer	20% for silicon [175] 300% for Cu-coated polypyrrole [176]
	Serpentine/fractal electrodes	Cu/Cr/polyimide on PDMS substrate	99% for parallel plate structure and 55% for circular involution structure [165]
	Origami/kirigami electrodes	PET film between graphene electrodes	8% in the X-axis and 15% in the Y-axis [49]
	Crack-based bilayer	Metal coating polymer bilayer	2% for Pt film with micro-/nanocracks [76]
Conductive nanocomposite	Wavy structure by prestretch	1. Silver nanowires in PDMS surface 2. Carbon nanomaterials, (i.e., CNTs & graphene films)	50% for AgNWs [177] 100% for CNTs nanoribbon composite [178]
Intrinsically stretchable conductive film	3D network structure	PEDOT:PSS film	100% strain with 4,100 S/cm conductivity [179]

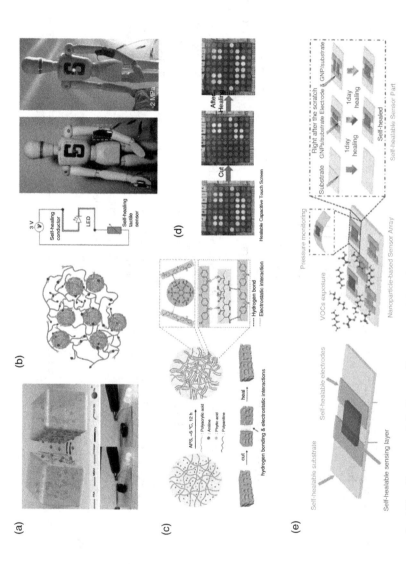

Figure 4.5 Progress of self-healing materials for tactile sensors. (a) Schematic illustration of intrinsic self-healing hydrogel *via* dynamic ionic interactions, and the LED response to applied pressure on the hydrogel in a complete circuit. Reproduced with permission from Ref. [180]. Copyright 2017 John Wiley and Sons. (b) Design and images of electrically and mechanically self-healing electronic sensor skin, and

Caption for Figure 4.5 (cont.)

their proposed interaction of oligomer chains with mNi particles. Reproduced with permission from Ref. [16]. Copyright 2012 Nature Publishing Group. (c) Synthetic processes of self-healable composite-based piezoresistive sensor, and self-healing mechanism due to hydrogen bonding and electrostatic interactions. Reproduced with permission from Ref. [181]. Copyright 2018 John Wiley and Sons. (d) Images of a healable transparent capacitive touch screen sensor based on a healable silver nanowire-polymer composite electrode. Reproduced with permission from Ref. [182]. Copyright 2014 American Chemical Society. (e) Design of a global self-healing pressure sensor system based on a chemiresistive transduction mechanism, incorporating self-healable polymer substrate, self-healable silver-polymer composite electrodes, and self-healable sensing films based on functionalized gold nanoparticles (GNPs). Reproduced with permission from Ref. [183], [184]. Copyright 2016 American Chemical Society and 2016 John Wiley and Sons.

[185], [186], [187]. As flexible electronics incorporate both soft and rigid materials, they can be damaged when severe forces are applied. When the external strain surpasses the ability of a device that can endure mechanical deformation, the device loses its function.

Furthermore, with the increasing complexity of integrated electronic devices, materials with self-healing properties have become the subject of intensive research. The aim is to create flexible sensors that can survive mechanical damage, recover from device failure, and maintain device functionality with minimal degradation. Self-healing has been extended to polymer material properties such as conductivity and transparency, and there are also some good summaries on the design of self-healing polymeric materials in electronics [188], [189]. Self-healing polymers can be divided into two categories: extrinsic self-healing polymers and intrinsic self-healing polymers. The first category is extrinsic self-healing, which utilizes microcapsules or vascular networks of curing agents that can be released on damage or break to bridge the fractured interfaces or damaged domains. Apart from microcapsule healing, other extrinsic self-healing methods that can realize the self-healing function include the vascular approach (1D, 2D, and 3D vascular systems) and hollow tube approach (discrete channels and interconnected networks).

The second category is intrinsic self-healing, which employs dynamic bonds that can autonomously regenerate after damage or dissociation. These dynamic bonds include dynamic covalent bonds [190], [191], [192], hydrogen bonds [193], [194], [195], metal-ligand coordination [196], [197], [198], hydrophobic interactions [199], [200], and ionic interactions [201], [202], [203], [204]. Intrinsic self-healing conductive polymers are very suitable for designing smart electronic materials and devices that can repair mechanical damage at the molecular or micrometer scale and recover mechanical strength at the macroscopic scale. In this section, we summarize the recent research on intrinsic self-healing conductive polymers and their application in self-healing tactile sensors and skin-like electronics.

Tee et al. first reported an electronic skin based on both electrically and mechanically self-healing composite that is flexion and pressure sensitive, just like human skin [16]. The material system incorporates a supramolecular polymeric hydrogen-bonding network, and chemically compatible micro-nickel (μNi) particles with surface nanostructures. Instead of strong covalent bonds, many weak hydrogen bonds contribute to the self-healing function of the supramolecular hydrogen-bonding network. In addition, a low glass transition temperature, which is below room temperature, enables the self-healing process to complete at room temperature environments. Figure 4.5(b) shows the self-healing tactile sensor mounted on the palm of a fully articulated wooden mannequin. The tactile sensor circuit modulates the LED's light intensity depending on the magnitude of the tactile force applied to the sensor.

Wang et al. also presented a ternary polymer composite of polyaniline, polyacrylic acid, and phytic acid that has excellent self-healing properties and a high stretchability [181]. Figure 4.5(c) illustrates the synthetic process for the ternary polymer composite and shows the interaction among polyaniline chains, polyacrylic, and phytic acid. Based on the self-healing mechanism due to hydrogen bonding and electrostatic interactions, they designed a pressure sensor with a micropatterned polymer film.

Darabi et al. introduced a mechanically and electrically self-healing hydrogel that is based on physically and chemically cross-linked networks of polymer chains [180], as shown in Figure 4.5(a). They employed PPy-grafted chitosan (DCh-PPy) and acrylic acid (AA) monomers to form poly(acrylic acid) (PAA) double by a two-step synthesis. The reversible ionic interactions between carboxylic groups of PAA and NH groups of PPy, and ferric ions devoted to an eye-catching autonomous self-healing performance.

Besides sensors, self-healing polymers can also be used to design conductive and stretchable electrodes. Li et al. used a self-healable silver nanowire-polymer composite electrode to design a healable, transparent, capacitive tactile screen sensor [182]. The composite electrode consists of a layer of silver nanowire percolation network that is located on the surface of the substrate. Figure 4.5(d) demonstrates the healing process of the touch screen sensor. Before cutting, a smiley face presents on the touch screen; and when cutting along the red dash line, only one-half of a smiley face is left; but after the healing process, the entire smiley face can be seen again. Moreover, Jin et al. designed a flexible, nanoparticle-based sensor array with a promising global self-healing property in detecting pressure variation [183], [184], as illustrated in Figure 4.5(e). The resultant tactile sensor is based on a chemiresistive transduction mechanism and incorporates three parts: the self-healable polymer substrate from self-healing disulfide-cross-linked polyurethane, self-healable silver-polymer composite electrodes, and self-healable sensing films based on functionalized gold nanoparticles (GNPs). The global self-healing sensors are expected to be used in future integrated wearable electronics and human health monitoring systems.

Self-healing sensors and systems are inherently robust to damage and can help with sustainability efforts and exploration of more challenging environments, such as degradation of polymer over time due to fatigue, and mechanical damage during operation [176]. Continued development of the electronic materials with self-healing capability will increase the chance that a fully self-healable electronic skin can be deployed in the near future for robotics applications.

4.4 Biocompatibility

The skin electronics devices demand biocompatibility due to the sensors directly mounted on human skin or organs. In particular, biocompatibility is a highly desirable requirement for implantable medical sensors interacting with living tissues. Although different biocompatible materials have been fabricated, most of these materials are insulators whereas electronic devices need full or partial conductivity. In the last decade, the key issues of flexible skin electronics are the mechanical mismatch of active film under stress/strain effect and the operation stability of electrical signals. However, minimization of invasiveness and skin irritation is the new challenge when the skin electronics is mounted on human skin.

For example, most flexible sensors are made on planar substrates, but long-time use would block the sweat, causing user discomfort. Miyamoto et al. in the Someya group designed gas-permeable, inflammation-free, and stretchable on-skin sensors [205], shown in Figure 4.6(a, b). The on-skin electronics system is based on a mesh structure made of biocompatible polyvinyl alcohol (PVA) with good gas permeability. When the nanomesh conductors are put on human skin and sprayed with water, the PVA nanofibers quickly dissolve and the nanomesh conductors adhere onto the skin. The researchers employed a one-week skin patch test and demonstrated that the risk of inflammation induced by on-skin electronics can be effectively inhibited by usage of the nanomesh conductors. In addition, the on-skin sensors can precisely measure electromyogram (EMG) recordings with minimal discomfort to the user.

The implantable sensors can provide continuous monitoring on tissue strain during rehabilitation protocols and during a patient's daily activities. Conventional implantable sensors, however, have limited biocompatibility or are designed for laboratory principle studies rather than clinical applications. Boutry et al. in the Bao group introduced an implantable and biodegradable strain and pressure sensor for orthopaedic application [24]. The design concept of real-time tendon healing assessment is illustrated in Figure 4.6(c). The sensor is manufactured to degrade when working lifetime is finished, without a second surgery to remove the device. The sensor is designed to be capable of successfully measuring the physiological strain signals on a real tendon. The in vivo experiment indicated that the stable operation of the sensor is more than 2 weeks, and the sensitivity is comparable to the reference sensor until the Mg electrodes start to degrade.

Choi et al. designed Ag-Au nanocomposites incorporating ultralong gold-coated silver nanowires in an elastomeric block-copolymer matrix [206]. The Ag-Au nanocomposite is fabricated by mixing Ag-Au nanowires, poly(styrene-butadiene-styrene) (SBS) elastomer, and an additive, hexylamine, in toluene,

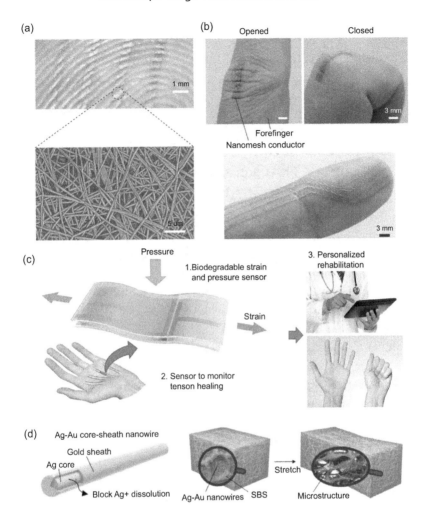

Figure 4.6 Progress of biocompatible skin electronics pressure sensor. (a) On-skin nanomesh electronics: photos of a nanomesh conductor attached on a fingertip with a high level of conformability, and a SEM image of a nanomesh conductor on a silicone skin replica by dissolving PVA nanofibers. (b) An on-skin wireless sensor system based on nanomesh conductors attached to a finger, maintaining conductance when the hand is opened and closed. Reproduced with permission from Ref. [205] Copyright 2017 Nature Publishing Group. (c) Schematic of a biodegradable and biocompatible pressure and strain sensor, which is adhered to a tendon for real-time healing measurement and to realize rehabilitation protocol after a tendon repair to be personalized for each patient. Reproduced with permission from Ref. [24] Copyright 2018 Nature Publishing Group. (d) Illustration of a microstructured Ag-Au nanocomposite made by

illustrated in Figure 4.6(d). The nanocomposites have optimized conductivity (41,850 S cm^{-1}) and stretchability of 266%. The outside thick gold sheath coated on the silver nanowire surface protects the internal silver from oxidation and ion leaching, which makes the nanocomposite biocompatible and conductive. On the basis of the nanocomposites, they fabricated one wearable and implantable compliant device, which is conformally integrated with human skin and swine heart for continuous electrophysiological recording.

4.5 Multifunction Electronic Skins

Recently, the field of neuroscience enabled significant gains in understanding how the human nervous systems work at such high performance yet in a power-efficient manner. Discoveries made by neuroscientists have already made an impact in computer science and electronics, in the form of artificial neural networks and neuromorphic hardware [208]–[210]. Not surprisingly, the growth in these fields has also encouraged materials engineers to design human-inspired sensors that can achieve the wide variety of sensing capabilities seen in humans. Indeed, there have been numerous research attempts to fabricate multifunctional sensors that can "express" the full range of human sensations.

For instance, Zhao et al. reported the use of stretchable optical waveguides for optoelectronic strain sensing in a prosthetic hand, which can be utilized as an integrated system of curvature, elongation, and pressure sensors, and their results indicated that the prosthetic hand can distinguish the shape and softness of three tomatoes and select the ripe one [19]. Figure 4.7(a) illustrates that a prosthetic hand with soft fingers identified differences in shape and roughness by touching and laterally scanning seven 3D-printed surfaces. After a simple calibration using an inclined plane with known height and angle, the height profiles of the seven surfaces can be easy to reconstruct, capable of differentiating curves as small as 5 m^{-1} and roughness on the order of 100 μm. Recently, Wang et al. proposed a self-powered, high-resolution, and highly sensitive tactile sensor with low power consumption for real-time trajectory mapping [18], as shown in Figure 4.7(b). They employed the self-powered, flexible, single-electrode triboelectric sensor matrix by PET film as a dielectric substrate and several layers of Ag film as electrodes. A PDMS film was spin-coated onto

---

Caption for Figure 4.6 (cont.)

combining a mixture of Ag-Au nanowires in elastomer matrix, and the Ag-Au nanocomposite states before and after stretching. Reproduced with permission from Ref. [206] Copyright 2018 Nature Publishing Group.

the top and subsequently cured, and it serves as the electrification layer to induce triboelectric charges in the contact process. Finally, the output electric signals result from the coupling effect between contact electrification and electrostatic induction, leading to a highly sensitive tactile sensor with no need of external electrical power.

A novel approach for sensing uses light for real-time visualization of pressure-sensing data. This would reduce the number of wires used for data telemetry and could become a useful technique for collecting information from sensors that are distributed over a large area. Tunable coloration/light emission has been demonstrated by combining optical elements with pressure sensors based on resistance and capacitance. Recently, Lee et al. reported a pressure-responsive electroluminescent (EL) display, which allows for both sensing and visualization of tactile pressure [17]. The conventional absorbance and reflective modes of coloration display elements suffer from low brightness, slow response time, and low light efficiency, but interactive EL displays are excellent owing to the strength of ultrathin and high color contrast and efficiency. Figure 4.7(c) depicts the device schematic and working principle of the EL pressure-sensing display, which can both sense and directly visualize pressure.

Additionally, Larson et al. presented an EL material with the ability to endure large uniaxial stretching and surface area variation during active light emission [33]. The resultant highly stretchable EL sheets are capable of being simultaneously subjected to stretching, folding, and rolling for use in stretchable displays that can be integrated into the skin of a soft robot, with dynamic coloration and sensory feedback from external and internal stimuli. Figure 4.7(e) shows an undulating gait produced by pressurizing the chambers in sequence along the length of the crawler, and a hyperelastic light-emitting capacitor (HLEC) that enables light emission and tactile sensing in a thin rubber sheet that stretches to >480% strain conforming to the end of a pencil. The HLEC array can provide dynamic coloration and the potential for feedback control, which might be useful in future epidermal electronics and robotics.

Hua et al. reported a skin-inspired highly stretchable and conformable matrix network (SCMN), which can sense temperature, in-plane strain, humidity, light, magnetic field, and proximity [20]. Figure 4.7(d) indicates the schematic layout of a SCMN-based integrated sensor array with eight functions. The authors presented a blueprint of a world in which billions of objects, with tightly integrated sensors, processors, and actuators, can sense multiple stimuli and react accordingly. This integrated system would allow humans to interact and communicate with their environment.

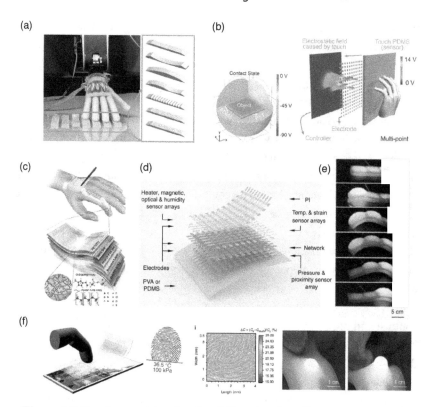

Figure 4.7 Development trends of multifunctional tactile sensor system. (a) Optoelectronically innervated soft prosthetic hand based on optoelectronic strain sensors by integrating curvature, stretch, and pressure sensors. Reproduced with permission from Ref. [19]. Copyright 2016 American Association for the Advancement of Science. (b) Schematic of a self-powered high-resolution and pressure-sensitive triboelectric sensor matrix for real-time tactile mapping. Reproduced with permission from Ref. [18]. Copyright 2016 John Wiley and Sons. (c) Structural design of a pressure-responsive electroluminescent (EL) display for both sensing and visualization of pressure. Reproduced with permission from Ref. [17]. Copyright 2018 American Chemical Society. (d) Schematic of a skin-inspired stretchable and conformable matrix networks for simultaneous multistimulus sensing. Reproduced with permission from Ref. [20]. Copyright 2018 Nature Publishing Group. (e) Variation of an undulating gait by pressurizing the chambers in a stretchable electroluminescent skin sensor for optical signaling and tactile sensing (right top), and images of the hyperelastic light-emitting capacitor (HLEC) conforming to the end of a pencil (right bottom). Reproduced with permission from Ref. [33]. Copyright 2016 American Association for the Advancement of Science. (f) Illustration of a transparent and flexible, capacitive fingerprint

Mechanosensation electronics is regarded as the core component of the system, and multifunctionalities are a critical technology to realize the human–machine interactions based on smart flexible electronics. An et al. designed a transparent and flexible capacitive fingerprint sensor array with multiplexed detection of tactile pressure and finger skin temperature [207]. They employed networks of hybrid nanostructures based on ultralong metal nanofibers and finer nanowires to fabricate flexible electrodes, which offers a capacitance change about 17 times better than that achieved using traditional ITO electrodes. Figure 4.7(f) depicts the schematic of the working principle for the multiplexed fingerprint sensor and the 2D mapping of relative changes in the capacitance of the fingerprint sensor on the touch of a human finger. The proposed multifunctional sensor array demonstrates the integration of pressure and temperature sensing, showing the potential for fingerprint recognition in enhancing security checks.

To make sensors more applicable in disease diagnoses, epidermal signals monitoring, and human–machine interfaces, wireless technologies are necessary for signal transmission from the sensor. By utilizing near-field electromagnetic coupling based on resonant frequencies between sensor devices and an external antenna, wireless detection and signal transmission can be achieved. Chen et al. developed the continuous wireless pressure monitoring system by designing small passive pressure sensors ($1 \times 1 \times 0.1$ mm^3) [211]. Their laminated sensor structures of substrate (polyimide), inductive spiral (Cu), and elastomer dielectric material were fabricated through a low-cost printing method. Styrene-butadiene-styrene (SBS) was selected as the dielectric material in the wireless sensor because of its low loss in the high-frequency range compared with other elastomers. SBS was cast to form micropyramidal structures for enhanced pressure sensitivity. When loads were applied, the change of coupling capacitance could shift down the resonant frequency. In addition, battery-free design of wireless systems and system dimension minimization technology were also explored [10], [212]. Wireless energy harvesting, power delivery, and communication strategies were developed for full-body distributed sensor arrays for pressure mapping. To attain skin-like stretchability, the stretchable antenna for far-field communication was studied [213]. Structural

---

Caption for Figure 4.7 (cont.)

sensor array with multiplexed, simultaneous detection of tactile pressure and finger skin temperature. Reproduced with permission from Ref. [207]. Copyright 2018 Nature Publishing Group.

engineering achieved the design by depositing metal films on flexible polymeric platforms.

5 Conclusions and Perspectives

In this Element, we discussed the recent progress of key components in mimicking human skin on artificial platforms via functional materials design, and sensor array architectures for large-area tactile sensors. The advances in materials development, namely, sensitivity, self-healing, stretchability, multifunction sensing, and the ability to integrate many different types of sensors, have created tremendous potential for e-skins to be used in prosthetics and human robotics.

Different sensing mechanisms were first discussed. Principles of pressure detection were studied and compared among various types of tactile sensors (e.g., capacitive, piezoresistive, optical, and so forth). Typical designs and materials used in these sensors were also addressed. For each type of sensor, the specific design of functional materials was studied and compared. Nanoscale materials, conductive layers, and microstructured polymers were rationally designed to provide highly sensitive detection of tactile pressure. Besides, structural design of these materials was also applied to provide new strategies for further improved sensing performance. For example, fractured graphene-coated sponge and interlocking of microstructures were presented for enhanced sensitivity and pressure detection range. A systematic study of parameterized materials and structure design of functional sensor materials was also discussed in this Element.

In addition to materials design, we provided an overview of several sensor hardware systems that have been used to demonstrate tactile sensors. Among these systems, matrix array implementations are the most common and are frequently used for applications that do not require high temporal resolutions. These arrays do not scale well but are straightforward to implement. As the matrix array gets larger, the temporal resolution degrades. To remedy this situation, the scalable architectures that we have examined partition taxels into modules that are connected to a bus/network. These modules relay sensing data using digital communication techniques. This systems-based approach is the way forward for designing real-world tactile sensing hardware that can scale up. And, although we do not discuss it here, the proliferation of AI techniques for processing large datasets could potentially affect hardware design decisions [79] and how these systems are made in the future.

Incorporation of multiple functions into tactile sensors will allow designers to use these sensors to create novel products. The ability to self-heal improves the reliability of sensors and could contribute to reduced maintenance costs. The increased adoption of wearables in fashion and real-time health monitoring has resulted in greater emphasis on sensors that have high stretchability and sensitivity. Based on these trends, we should also see flexible and stretchable electronics play an increasingly important role in enabling these applications. Additionally, advances in microstructure engineering will lead to improvements in the manufacturing of large-area tactile sensor arrays.

In conclusion, we anticipate that electronic skins will continue to use human skin as a template for design. Strategies to scale up the sensor density and achieve real-time feedback with high temporal resolution will allow roboticists to improve the ability of robots to interact with their environment and, importantly, with humans. We believe the future is extremely bright for the continued development of large-area, conformable, tactile sensor-based electronic skins as a critical component in the rapidly advancing robotics era.

References

[1] M. Ha, S. Lim, J. Park, D. S. Um, Y. Lee, and H. Ko, "Bioinspired interlocked and hierarchical design of ZnO nanowire arrays for static and dynamic pressure-sensitive electronic skins," *Adv. Funct. Mater.*, vol. 25, no. 19, pp. 2841–2849, 2015.

[2] Y. Kim *et al.*, "A bioinspired flexible organic artificial afferent nerve," *Science (80–.).*, vol. 360, no. 6392, pp. 998–1003, Jun. 2018.

[3] D. Silvera-Tawil, D. Rye, and M. Velonaki, "Artificial skin and tactile sensing for socially interactive robots: A review," *Rob. Auton. Syst.*, vol. 63, no. P3, pp. 230–243, 2015.

[4] J. A. Rogers, T. Someya, and Y. Huang, "Materials and mechanics for stretchable electronics," *Science (80–.).*, vol. 327, no. 5973, pp. 1603–1607, Mar. 2010.

[5] T. Sekitani and T. Someya, "Stretchable, large-area organic electronics," *Adv. Mater.*, vol. 22, no. 20, pp. 2228–2246, 2010.

[6] S. Park, M. Vosguerichian, and Z. Bao, "A review of fabrication and applications of carbon nanotube film-based flexible electronics," *Nanoscale*, vol. 5, no. 5, pp. 1727–1752, 2013.

[7] B. C.-K. Tee *et al.*, "A skin-inspired organic digital mechanoreceptor," *Science (80–.).*, vol. 350, no. 6258, pp. 313–316, Oct. 2015.

[8] T. Someya, T. Sekitani, S. Iba, Y. Kato, H. Kawaguchi, and T. Sakurai, "A large-area, flexible pressure sensor matrix with organic field-effect transistors for artificial skin applications," *Proc. Natl. Acad. Sci.*, vol. 101, no. 27, pp. 9966–9970, 2004.

[9] X. Xiao *et al.*, "High-strain sensors based on ZnO nanowire/polystyrene hybridized flexible films," *Adv. Mater.*, vol. 23, no. 45, pp. 5440–5444, 2011.

[10] J. Kim *et al.*, "Miniaturized battery-free wireless systems for wearable pulse oximetry," *Adv. Funct. Mater.*, vol. 27, no. 1, pp. 1–8, 2017.

[11] C. Pan *et al.*, "High-resolution electroluminescent imaging of pressure distribution using a piezoelectric nanowire LED array," *Nat. Photonics*, vol. 7, no. 9, pp. 752–758, 2013.

[12] G. Schwartz *et al.*, "Flexible polymer transistors with high pressure sensitivity for application in electronic skin and health monitoring," *Nat. Commun.*, vol. 4, no. May, p. 1859, 2013.

[13] S. Wang *et al.*, "Skin electronics from scalable fabrication of an intrinsically stretchable transistor array," *Nature*, vol. 555, no. 7694, pp. 83–88, 2018.

[14] J. Byun *et al.*, "Electronic skins for soft, compact, reversible assembly of wirelessly activated fully soft robots," *Sci. Robot.*, vol. 3, no. 18, p. eaas9020, 2018.

[15] J. Xu *et al.*, "Highly stretchable polymer semiconductor films through the nanoconfinement effect," *Science (80–.)*., vol. 355, no. 6320, p. 59 LP-64, Jan. 2017.

[16] B. C. K. Tee, C. Wang, R. Allen, and Z. Bao, "An electrically and mechanically self-healing composite with pressure- and flexion-sensitive properties for electronic skin applications," *Nat. Nanotechnol.*, vol. 7, no. 12, pp. 825–832, 2012.

[17] S. W. Lee *et al.*, "Electroluminescent pressure-sensing displays," *ACS Appl. Mater. Interfaces*, vol. 10, no. 16, pp. 13757–13766, 2018.

[18] X. Wang *et al.*, "Self-powered high-resolution and pressure-sensitive triboelectric sensor matrix for real-time tactile mapping," *Adv. Mater.*, vol. 28, no. 15, pp. 2896–2903, 2016.

[19] H. Zhao, K. O'Brien, S. Li, and R. F. Shepherd, "Optoelectronically innervated soft prosthetic hand via stretchable optical waveguides," *Sci. Robot.*, vol. 1, no. 1, p. eaai7529, 2016.

[20] Q. Hua *et al.*, "Skin-inspired highly stretchable and conformable matrix networks for multifunctional sensing," *Nat. Commun.*, vol. 9, no. 1, pp. 1–12, 2018.

[21] U. Hagn *et al.*, "DLR MiroSurge: A versatile system for research in endoscopic telesurgery," *Int. J. Comput. Assist. Radiol. Surg.*, vol. 5, no. 2, pp. 183–193, 2010.

[22] H. Jang, Y. J. Park, X. Chen, T. Das, M. S. Kim, and J. H. Ahn, "Graphene-based flexible and stretchable electronics," *Adv. Mater.*, vol. 28, no. 22, pp. 4184–4202, 2016.

[23] Y. Qian *et al.*, "Stretchable organic semiconductor devices," *Adv. Mater.*, vol. 28, no. 42, pp. 9243–9265, 2016.

[24] C. M. Boutry *et al.*, "A stretchable and biodegradable strain and pressure sensor for orthopaedic application," *Nat. Electron.*, vol. 1, no. 5, pp. 314–321, 2018.

[25] S. J. Benight, C. Wang, J. B. H. Tok, and Z. Bao, "Stretchable and self-healing polymers and devices for electronic skin," *Prog. Polym. Sci.*, vol. 38, no. 12, pp. 1961–1977, 2013.

[26] J. Y. Oh *et al.*, "Intrinsically stretchable and healable semiconducting polymer for organic transistors," *Nature*, vol. 539, no. 7629, pp. 411–415, 2016.

[27] G. Zhu *et al.*, "Self-powered, ultrasensitive, flexible tactile sensors based on contact electrification," *Nano Lett.*, vol. 14, no. 6, pp. 3208–3213, 2014.

[28] C. G. Núñez, W. T. Navaraj, E. O. Polat, and R. Dahiya, "Energy-autonomous, flexible, and transparent tactile skin," *Adv. Funct. Mater.*, vol. 27, no. 18, 2017.

[29] L. Manjakkal, C. G. Núñez, W. Dang, and R. Dahiya, "Flexible self-charging supercapacitor based on graphene-Ag-3D graphene foam electrodes," *Nano Energy*, vol. 51, no. June, pp. 604–612, 2018.

[30] C. M. Boutry, A. Nguyen, Q. O. Lawal, A. Chortos, S. Rondeau-Gagné, and Z. Bao, "A sensitive and biodegradable pressure sensor array for cardiovascular monitoring," *Adv. Mater.*, vol. 27, no. 43, pp. 6954–6961, 2015.

[31] V. R. Feig, H. Tran, and Z. Bao, "Biodegradable polymeric materials in degradable electronic devices," *ACS Cent. Sci.*, vol. 4, no. 3, pp. 337–348, 2018.

[32] T. Sekitani *et al.*, "Stretchable active-matrix organic light-emitting diode display using printable elastic conductors," *Nat. Mater.*, vol. 8, no. 6, pp. 494–499, 2009.

[33] C. Larson *et al.*, "Highly stretchable electroluminescent skin for optical signaling and tactile sensing," *Science (80–.)*, vol. 351, no. 6277, pp. 1071–1074, 2016.

[34] M. L. Hammock, A. Chortos, B. C. K. Tee, J. B. H. Tok, and Z. Bao, "25th anniversary article: The evolution of electronic skin (E-Skin): A brief history, design considerations, and recent progress," *Adv. Mater.*, vol. 25, no. 42, pp. 5997–6038, 2013.

[35] M. Amjadi, K. U. Kyung, I. Park, and M. Sitti, "Stretchable, skin-mountable, and wearable strain sensors and their potential applications: a review," *Adv. Funct. Mater.*, vol. 26, no. 11, pp. 1678–1698, 2016.

[36] J. Heikenfeld *et al.*, "Wearable sensors: Modalities, challenges, and prospects," *Lab Chip*, vol. 18, no. 2, pp. 217–248, 2018.

[37] R. S. Dahiya, M. Valle, and G. Metta, "System approach: A paradigm for robotic tactile sensing," *Int. Work. Adv. Motion Control. AMC*, vol. 1, pp. 110–115, 2008.

[38] R. S. Dahiya, P. Mittendorfer, M. Valle, G. Cheng, and V. J. Lumelsky, "Directions toward effective utilization of tactile skin: A review," *IEEE Sens. J.*, vol. 13, no. 11, pp. 4121–4138, 2013.

[39] R. S. Dahiya, G. Metta, M. Valle, and G. Sandini, "Tactile sensing – from humans to humanoids," *IEEE Trans. Robot.*, vol. 26, no. 1, pp. 1–20, 2010.

[40] A. Aijaz and M. Sooriyabandara, "The tactile internet for industries: A review," *Proc. IEEE*, vol. 107, no. 2, pp. 414–435, 2019.

[41] R. S. Dahiya and M. Valle, *Robotic Tactile Sensing*. 2012.

[42] S. C. B. Mannsfeld *et al.*, "Highly sensitive flexible pressure sensors with microstructured rubber dielectric layers," *Nat. Mater.*, vol. 9, no. 10, pp. 859–864, 2010.

[43] S. Y. Kim, S. Park, H. W. Park, D. H. Park, Y. Jeong, and D. H. Kim, "Highly sensitive and multimodal all-carbon skin sensors capable of simultaneously detecting tactile and biological stimuli," *Adv. Mater.*, vol. 27, no. 28, pp. 4178–4185, 2015.

[44] B. Nie, R. Li, J. Cao, J. D. Brandt, and T. Pan, "Flexible transparent iontronic film for interfacial capacitive pressure sensing," *Adv. Mater.*, vol. 27, no. 39, pp. 6055–6062, 2015.

[45] J. Li *et al.*, "Healable capacitive touch screen sensors based on transparent composite electrodes comprising silver nanowires and a furan/maleimide diels-Alder cycloaddition polymer," *ACS Nano*, vol. 8, no. 12, pp. 12874–12882, 2014.

[46] C. Bartolozzi, L. Natale, F. Nori, and G. Metta, "Robots with a sense of touch," *Nat. Mater.*, vol. 15, no. 9, pp. 921–925, 2016.

[47] A. P. Gerratt, H. O. Michaud, and S. P. Lacour, "Elastomeric electronic skin for prosthetic tactile sensation," *Adv. Funct. Mater.*, vol. 25, no. 15, pp. 2287–2295, 2015.

[48] J. W. Jeong *et al.*, "Capacitive epidermal electronics for electrically safe, long-term electrophysiological measurements," *Adv. Healthc. Mater.*, vol. 3, no. 5, pp. 642–648, 2014.

[49] M. Kang, J. Kim, B. Jang, Y. Chae, J. H. Kim, and J. H. Ahn, "Graphene-based three-dimensional capacitive touch sensor for wearable electronics," *ACS Nano*, vol. 11, no. 8, pp. 7950–7957, 2017.

[50] D. J. Lipomi *et al.*, "Skin-like pressure and strain sensors based on transparent elastic films of carbon nanotubes," *Nat. Nanotechnol.*, vol. 6, no. 12, pp. 788–792, 2011.

[51] S. Yao and Y. Zhu, "Wearable multifunctional sensors using printed stretchable conductors made of silver nanowires," *Nanoscale*, vol. 6, no. 4, pp. 2345–2352, 2014.

[52] J. Lee *et al.*, "Conductive fiber-based ultrasensitive textile pressure sensor for wearable electronics," *Adv. Mater.*, vol. 27, no. 15, pp. 2433–2439, 2015.

[53] S. Gong *et al.*, "A wearable and highly sensitive pressure sensor with ultrathin gold nanowires," *Nat. Commun.*, vol. 5, 2014.

[54] B. C. K. Tee *et al.*, "A skin-inspired organic digital mechanoreceptor," *Science (80-.).*, vol. 350, no. 6258, pp. 313–316, 2015.

[55] H. Tian *et al.*, "A graphene-based resistive pressure sensor with record-high sensitivity in a wide pressure range," *Sci. Rep.*, vol. 5, pp. 1–6, 2015.

[56] L. Wang *et al.*, "PDMS/MWCNT-based tactile sensor array with coplanar electrodes for crosstalk suppression," *Microsystems Nanoeng.*, vol. 2, no. March, p. 16065, 2016.

[57] K. Takei, Z. Yu, M. Zheng, H. Ota, T. Takahashi, and A. Javey, "Highly sensitive electronic whiskers based on patterned carbon nanotube and silver nanoparticle composite films," *Proc. Natl. Acad. Sci.*, vol. 111, no. 5, pp. 1703–1707, 2014.

[58] X. Li *et al.*, "Stretchable and highly sensitive graphene-on-polymer strain sensors," *Sci. Rep.*, vol. 2, pp. 1–6, 2012.

[59] B. Zhu *et al.*, "Microstructured graphene arrays for highly sensitive flexible tactile sensors," *Small*, vol. 10, no. 18, pp. 3625–3631, 2014.

[60] Z. Lou, S. Chen, L. Wang, K. Jiang, and G. Shen, "An ultra-sensitive and rapid response speed graphene pressure sensors for electronic skin and health monitoring," *Nano Energy*, vol. 23, pp. 7–14, 2016.

[61] C. L. Choong *et al.*, "Highly stretchable resistive pressure sensors using a conductive elastomeric composite on a micropyramid array," *Adv. Mater.*, vol. 26, no. 21, pp. 3451–3458, 2014.

[62] R. S. Karmakar *et al.*, "Cross-talk immunity of PEDOT:PSS pressure sensing arrays with gold nanoparticle incorporation," *Sci. Rep.*, vol. 7, no. 1, pp. 1–10, 2017.

[63] C. Dagdeviren *et al.*, "Conformable amplified lead zirconate titanate sensors with enhanced piezoelectric response for cutaneous pressure monitoring," *Nat. Commun.*, vol. 5, pp. 1–10, 2014.

[64] L. Persano *et al.*, "High performance piezoelectric devices based on aligned arrays of nanofibers of poly(vinylidenefluoride-co-trifluoroethylene)," *Nat. Commun.*, vol. 4, pp. 1610–1633, 2013.

[65] Y. Jeong *et al.*, "Psychological tactile sensor structure based on piezoelectric nanowire cell arrays," *RSC Adv.*, vol. 5, no. 50, pp. 40363–40368, 2015.

[66] Y. Lee *et al.*, "Flexible ferroelectric sensors with ultrahigh pressure sensitivity and linear response over exceptionally broad pressure range," *ACS Nano*, vol. 12, no. 4, pp. 4045–4054, 2018.

[67] J. H. Lee *et al.*, "Micropatterned P(VDF-TrFE) film-based piezoelectric nanogenerators for highly sensitive self-powered pressure sensors," *Adv. Funct. Mater.*, vol. 25, no. 21, pp. 3203–3209, 2015.

[68] X. Chen *et al.*, "High-performance piezoelectric nanogenerators with imprinted P(VDF-TrFE)/BaTiO3 nanocomposite micropillars for self-powered flexible sensors," *Small*, vol. 13, no. 23, pp. 1–12, 2017.

[69] X. Wang *et al.*, "Full dynamic-range pressure sensor matrix based on optical and electrical dual-mode sensing," *Adv. Mater.*, vol. 29, no. 15, 2017.

[70] X. Pu *et al.*, "Ultrastretchable, transparent triboelectric nanogenerator as electronic skin for biomechanical energy harvesting and tactile sensing," *Sci. Adv.*, vol. 3, no. 5, pp. 1–11, 2017.

[71] X. Wang *et al.*, "A highly stretchable transparent self-powered triboelectric tactile sensor with metallized nanofibers for wearable electronics," *Adv. Mater.*, vol. 30, no. 12, pp. 1–8, 2018.

[72] Z. L. Wang, J. Chen, and L. Lin, "Progress in triboelectric nanogenerators as a new energy technology and self-powered sensors," *Energy Environ. Sci.*, vol. 8, no. 8, pp. 2250–2282, 2015.

[73] M. Ramuz, B. C. K. Tee, J. B. H. Tok, and Z. Bao, "Transparent, optical, pressure-sensitive artificial skin for large-area stretchable electronics," *Adv. Mater.*, vol. 24, no. 24, pp. 3223–3227, 2012.

[74] C. Wang *et al.*, "User-interactive electronic skin for instantaneous pressure visualization," *Nat. Mater.*, vol. 12, no. 10, pp. 899–904, 2013.

[75] T. Someya *et al.*, "Conformable, flexible, large-area networks of pressure and thermal sensors with organic transistor active matrixes," *Proc. Natl. Acad. Sci.*, vol. 102, no. 35, pp. 12321–12325, 2005.

[76] D. Kang *et al.*, "Ultrasensitive mechanical crack-based sensor inspired by the spider sensory system," *Nature*, vol. 516, no. 7530, pp. 222–226, 2014.

[77] M. Lee and H. Nicholls, "Review Article: Tactile sensing for mechatronics – a state of the art survey," *Mechatronics*, vol. 9, no. 1, pp. 1–31, 1999.

[78] J. Dargahi and S. Najarian, "Human tactile perception as a standard for artificial tactile sensing – a review.," *Int. J. Med. Robot.*, vol. 1, no. 1, pp. 23–35, 2004.

[79] W. T. Navaraj *et al.*, "Nanowire FET based neural element for robotic tactile sensing skin," *Front. Neurosci.*, vol. 11, no. SEP, p. 501, 2017.

[80] R. S. Johansson and A. B. Vallbo, "Tactile sensibility in the human hand: relative and absolute densities of four types of mechanoreceptive units in glabrous skin.," *J. Physiol.*, vol. 286, no. 1, pp. 283–300, 1979.

[81] M. Boniol, J. P. Verriest, R. Pedeux, and J. F. Doré, "Proportion of skin surface area of children and young adults from 2 to 18 years old," *J. Invest. Dermatol.*, vol. 128, no. 2, pp. 461–464, 2008.

[82] F. Mancini *et al.*, "Whole-body mapping of spatial acuity for pain and touch," *Ann. Neurol.*, vol. 75, no. 6, pp. 917–924, 2014.

[83] A. Chortos, J. Liu, and Z. Bao, "Pursuing prosthetic electronic skin," *Nat. Mater.*, vol. 15, no. 9, pp. 937–950, 2016.

[84] E. L. Mackevicius, M. D. Best, H. P. Saal, and S. J. Bensmaia, "Millisecond precision spike timing shapes tactile perception," *J. Neurosci.*, vol. 32, no. 44, pp. 15309–15317, 2012.

[85] R. S. Johansson and I. Birznieks, "First spikes in ensembles of human tactile afferents code complex spatial fingertip events," *Nat. Neurosci.*, vol. 7, no. 2, pp. 170–177, 2004.

[86] W. W. Lee, S. L. Kukreja, and N. V. Thakor, "Discrimination of dynamic tactile contact by temporally precise event sensing in spiking neuromorphic networks," *Front. Neurosci.*, vol. 11, no. JAN, pp. 1–14, 2017.

[87] X. Hu and W. Yang, "Planar capacitive sensors - Designs and applications," *Sens. Rev.*, vol. 30, no. 1, pp. 24–39, 2010.

[88] Y. J. Yang *et al.*, "An integrated flexible temperature and tactile sensing array using PI-copper films," *Int. J. Adv. Manuf. Technol.*, vol. 46, no. 9–12, pp. 945–956, 2010.

[89] W. Y. Chang, T. H. Fang, S. H. Yeh, and Y. C. Lin, "Flexible electronics sensors for tactile multi-touching," *Sensors*, vol. 9, no. 2, pp. 1188–1203, 2009.

[90] D. Elastomers, "Product Selection Guide," no. 3, pp. 1–4.

[91] C. C. Enz and G. C. Temes, "Circuit techniques for reducing the effects of op-amp imperfections: autozeroing, correlated double sampling, and chopper stabilization," *Proc. IEEE*, vol. 84, no. 1, pp. 1584–1614, 1996.

[92] M. Park, B.-G. Bok, J.-H. Ahn, and M.-S. Kim, "Recent advances in tactile sensing technology," *Micromachines*, vol. 9, no. 7, p. 321, 2018.

[93] R. S. Johansson and J. R. Flanagan, "Coding and use of tactile signals from the fingertips in object manipulation tasks," *Nat. Rev. Neurosci.*, vol. 10, no. 5, pp. 345–359, 2009.

[94] S. Actuators, "Wire harness," *Assembly Magazine*, 1981. [Online]. Available: www.assemblymag.com/articles/92263-wire-harness-recyc ling. [Accessed: 29-Aug-2018].

[95] R. S. Dahiya, G. Metta, and M. Valle, "Development of fingertip tactile sensing chips for humanoid robots," *IEEE 2009 Int. Conf. Mechatronics, ICM 2009*, vol. 00, no. April, pp. 1–6, 2009.

[96] W. W. Lee, S. L. Kukreja, and N. V. Thakor, "Live demonstration: A kilohertz kilotaxel tactile sensor array for investigating spatiotemporal features in neuromorphic touch," in *IEEE Biomedical Circuits and Systems Conference: Engineering for Healthy Minds and Able Bodies, BioCAS 2015 - Proceedings*, 2015.

[97] K. Takei *et al.*, "Nanowire active-matrix circuitry for low-voltage macroscale artificial skin," *Nat. Mater.*, vol. 9, no. 10, pp. 821–826, 2010.

[98] C. Yeom, K. Chen, D. Kiriya, Z. Yu, G. Cho, and A. Javey, "Large-area compliant tactile sensors using printed carbon nanotube active-matrix backplanes," *Adv. Mater.*, vol. 27, no. 9, pp. 1561–1566, 2015.

[99] Y. Zang, F. Zhang, D. Huang, X. Gao, C. A. Di, and D. Zhu, "Flexible suspended gate organic thin-film transistors for ultra-sensitive pressure detection," *Nat. Commun.*, vol. 6, pp. 1–9, 2015.

[100] W. Wu, X. Wen, and Z. L. Wang, "Taxel-addressable matrix of vertical-nanowire piezotronic transistors for active and adaptive tactile imaging," *Science (80–.).*, vol. 340, no. 6135, pp. 952–957, May 2013.

[101] L. Barboni, R. S. Dahiya, G. Metta, and M. Valle, "Interface electronics design for POSFET devices based tactile sensing systems," *Proc. - IEEE Int. Work. Robot Hum. Interact. Commun.*, pp. 686–690, 2010.

[102] M. Sergio, N. Manaresi, M. Tartagni, R. Guerrieri, and R. Canegallo, "A textile based capacitive pressure sensor," *Proc. IEEE Sensors*, vol. 2, pp. 1625–1630.

[103] G. Walker, "Fundamentals of projected-capacitive touch technology," *Tech. Present.*, 2014.

[104] PPS, "Industrial TactArray Sensors," *Sensors (Peterborough, NH)*, pp. 90045–90045, 2000.

[105] F. Vidal-Verdú *et al.*, "A large area tactile sensor patch based on commercial force sensors," *Sensors*, vol. 11, no. 5, pp. 5489–5507, 2011.

[106] T. D'Alessio, "Measurement errors in the scanning of piezoresistive sensors arrays," *Sensors Actuators, A Phys.*, vol. 72, no. 1, pp. 71–76, 1999.

[107] R. S. Dahiya, L. Lorenzelli, G. Metta, and M. Valle, "POSFET devices based tactile sensing arrays," *ISCAS 2010-2010 IEEE Int. Symp. Circuits Syst. Nano-Bio Circuit Fabr. Syst.*, pp. 893–896, 2010.

[108] J. Ulmen and M. Cutkosky, "A robust, low-cost and low-noise artificial skin for human-friendly robots," *Proc. - IEEE Int. Conf. Robot. Autom.*, pp. 4836–4841, 2010.

[109] G. Cannata, M. Maggiali, G. Metta, and G. Sandini, "An embedded artificial skin for humanoid robots," *IEEE Int. Conf. Multisens. Fusion Integr. Intell. Syst.*, pp. 434–438, 2008.

[110] F. Bergner, P. Mittendorfer, E. Dean-Leon, and G. Cheng, "Event-based signaling for reducing required data rates and processing power in a large-scale artificial robotic skin," *IEEE Int. Conf. Intell. Robot. Syst.*, vol. 2015–December, pp. 2124–2129, 2015.

[111] M. Makihata *et al.*, "A 1.7mm^3 MEMS-on-CMOS tactile sensor using human-inspired autonomous common bus communication," *2013 Transducers Eurosensors XXVII 17th Int. Conf. Solid-State Sensors, Actuators Microsystems, TRANSDUCERS EUROSENSORS 2013*, no. June, pp. 2729–2732, 2013.

[112] S. Asano, M. Muroyama, T. Nakayama, Y. Hata, Y. Nonomura, and S. Tanaka, "3-axis fully-integrated capacitive tactile sensor with flip-bonded CMOS on LTCC interposer," *Sensors (Switzerland)*, vol. 17, no. 11, pp. 1–14, 2017.

[113] C. Shao *et al.*, "A tactile sensor network system using a multiple sensor platform with a dedicated CMOS-LSI for robot applications," *Sensors*, vol. 17, no. 9, p. 1974, 2017.

[114] C. Bartolozzi *et al.*, "Event-driven encoding of off-the-shelf tactile sensors for compression and latency optimisation for robotic skin," *IEEE Int. Conf. Intell. Robot. Syst.*, vol. 2017–September, pp. 166–173, 2017.

[115] Y. Afsar, T. Moy, N. Brady, S. Wagner, J. C. Sturm, and N. Verma, "Large-scale acquisition of large-area sensors using an array of frequency-hopping ZnO thin-film-transistor oscillators," *Dig. Tech. Pap. - IEEE Int. Solid-State Circuits Conf.*, vol. 60, pp. 256–257, 2017.

[116] Y. Afsar, T. Moy, N. Brady, S. Wagner, J. C. Sturm, and N. Verma, "An architecture for large-area sensor acquisition using frequency-hopping ZnO TFT DCOs," *IEEE J. Solid-State Circuits*, vol. 53, no. 1, pp. 297–308, 2018.

[117] R. Pallàs-Areny, *Sensors and Signal Conditioning*, 2nd ed. Wiley-Interscience, 2003.

[118] Jacob Fraden, *Handbook of Modern Sensors: PHYSICS, DESIGNS, and APPLICATIONS*, 3rd ed. Springer, 2004.

[119] W. Kester, "High impedance sensors," in *Practical Design Techniques for Sensor Signal Conditioning*, Analog Devices, 1999.

[120] L. Barboni, M. Valle, and G. Carlini, "Smart readout design for tactile sensing devices," *2011 18th IEEE Int. Conf. Electron. Circuits, Syst. ICECS 2011*, pp. 476–479, 2011.

[121] W. Kester, J. Bryant, W. Jung, S. Wurcer, and C. Kitchin, *Sensor Signal Conditioning*. 2004.

[122] L. Seminara, L. Pinna, M. Capurro, and M. Valle, "A tactile sensing system based on arrays of piezoelectric polymer transducers," in *Smart Actuation and Sensing Systems – Recent Advances and Future Challenges*, 2012, pp. 611–638.

[123] J. O'Dowd, A. Callanan, G. Banarie, and E. Company-Bosch, "Capacitive sensor interfacing using sigma-delta techniques," *Proc. IEEE Sensors*, vol. 2005, pp. 951–954, 2005.

[124] V. Ferrari, C. Ghidini, D. Marioli, and A. Taroni, "Oscillator-based signal conditioning for resistive sensors," *Conf. Rec. - IEEE Instrum. Meas. Technol. Conf.*, vol. 2, pp. 1490–1494, 1997.

[125] R. Schreier and G. C. Temes, *Understanding Delta-Sigma Data Converters*, 2nd ed. Wiley-IEEE Press, 2005.

[126] P. Madaan and P. Kaur, "Capacitive sensing made easy, Part 1: An introduction to different capacitive sensing technologies," *EE Times Name*, no. April, pp. 1–8, 2012.

[127] R. Martins, N. Lourenço, and N. Horta, *Analog Integrated Circuit Design Automation*, 1st ed. Wiley, 2017.

[128] J. Engel, N. Chen, C. Tucker, C. Liu, S. H. Kim, and D. Jones, "Flexible multimodal tactile sensing system for object identification," *Proc. IEEE Sensors*, pp. 563–566, 2006.

[129] J. G. Da Silva, A. A. De Carvalho, and D. D. Da Silva, "A strain gauge tactile sensor for finger-mounted applications," *IEEE Trans. Instrum. Meas.*, vol. 51, no. 1, pp. 18–22, 2002.

[130] J. Il Lee, M. G. Kim, M. Shikida, and K. Sato, "A table-shaped tactile sensor for detecting triaxial force on the basis of strain distribution," *Sensors (Switzerland)*, vol. 13, no. 12, pp. 16347–16359, 2013.

[131] W. Kester, "Chapter 2: Bridge Circuits," in *Practical Design Techniques for Sensor Signal Conditioning*, Analog Devices, 1999.

[132] H. Bin Yao *et al.*, "A flexible and highly pressure-sensitive graphene-polyurethane sponge based on fractured microstructure design," *Adv. Mater.*, vol. 25, no. 46, pp. 6692–6698, 2013.

[133] C. Pang *et al.*, "A flexible and highly sensitive strain-gauge sensor using reversible interlocking of nanofibres," *Nat. Mater.*, vol. 11, no. 9, pp. 795–801, 2012.

[134] G. Y. Bae *et al.*, "Linearly and highly pressure-sensitive electronic skin based on a bioinspired hierarchical structural array," *Adv. Mater.*, vol. 28, no. 26, pp. 5300–5306, 2016.

[135] K. Y. Chun *et al.*, "Highly conductive, printable and stretchable composite films of carbon nanotubes and silver," *Nat. Nanotechnol.*, vol. 5, no. 12, pp. 853–857, 2010.

[136] Y. Pang *et al.*, "Flexible, highly sensitive, and wearable pressure and strain sensors with graphene porous network structure," *ACS Appl. Mater. Interfaces*, vol. 8, no. 40, pp. 26458–26462, 2016.

[137] L. Sheng *et al.*, "Bubble-decorated honeycomb-like graphene film as ultrahigh sensitivity pressure sensors," *Adv. Funct. Mater.*, vol. 25, no. 41, pp. 6545–6551, 2015.

[138] S. Lee *et al.*, "A transparent bending-insensitive pressure sensor," *Nat. Nanotechnol.*, vol. 11, no. 5, pp. 472–478, 2016.

[139] I. D. Johnston, D. K. McCluskey, C. K. L. Tan, and M. C. Tracey, "Mechanical characterization of bulk Sylgard 184 for microfluidics and

microengineering," *J. Micromechanics Microengineering*, vol. 24, no. 3, 2014.

[140] K. Khanafer, A. Duprey, M. Schlicht, and R. Berguer, "Effects of strain rate, mixing ratio, and stress-strain definition on the mechanical behavior of the polydimethylsiloxane (PDMS) material as related to its biological applications," *Biomed. Microdevices*, vol. 11, no. 2, pp. 503–508, 2009.

[141] R. N. Palchesko, L. Zhang, Y. Sun, and A. W. Feinberg, "Development of polydimethylsiloxane substrates with tunable elastic modulus to study cell mechanobiology in muscle and nerve," *PLoS One*, vol. 7, no. 12, 2012.

[142] F. Schneider, T. Fellner, J. Wilde, and U. Wallrabe, "Mechanical properties of silicones for MEMS," *J. Micromechanics Microengineering*, vol. 18, no. 6, 2008.

[143] J. Park *et al.*, "Tailoring force sensitivity and selectivity by microstructure engineering of multidirectional electronic skins," *NPG Asia Mater.*, vol. 10, no. 4, pp. 163–176, 2018.

[144] B. C. K. Tee, A. Chortos, R. R. Dunn, G. Schwartz, E. Eason, and Z. Bao, "Tunable flexible pressure sensors using microstructured elastomer geometries for intuitive electronics," *Adv. Funct. Mater.*, vol. 24, no. 34, pp. 5427–5434, 2014.

[145] C. Pang *et al.*, "Highly skin-conformal microhairy sensor for pulse signal amplification," *Adv. Mater.*, vol. 27, no. 4, pp. 634–640, 2015.

[146] B. Su, S. Gong, Z. Ma, L. W. Yap, and W. Cheng, "Mimosa-inspired design of a flexible pressure sensor with touch sensitivity," *Small*, vol. 11, no. 16, pp. 1886–1891, 2015.

[147] T. Li *et al.*, "Flexible capacitive tactile sensor based on micropatterned dielectric layer," *Small*, vol. 12, no. 36, pp. 5042–5048, 2016.

[148] S. G. Yoon, B. J. Park, and S. T. Chang, "Highly sensitive piezocapacitive sensor for detecting static and dynamic pressure using ion-gel thin films and conductive elastomeric composites," *ACS Appl. Mater. Interfaces*, vol. 9, no. 41, pp. 36206–36219, 2017.

[149] X. Wang, Y. Gu, Z. Xiong, Z. Cui, and T. Zhang, "Silk-molded flexible, ultrasensitive, and highly stable electronic skin for monitoring human physiological signals," *Adv. Mater.*, vol. 26, no. 9, pp. 1336–1342, 2014.

[150] H. Park *et al.*, "Stretchable array of highly sensitive pressure sensors consisting of polyaniline nanofibers and Au-coated polydimethylsiloxane micropillars," *ACS Nano*, vol. 9, no. 10, pp. 9974–9985, 2015.

[151] L. Pan *et al.*, "An ultra-sensitive resistive pressure sensor based on hollow-sphere microstructure induced elasticity in conducting polymer film," *Nat. Commun.*, vol. 5, 2014.

[152] S. Chen *et al.*, "Noncontact heartbeat and respiration monitoring based on a hollow microstructured self-powered pressure sensor," *ACS Appl. Mater. Interfaces*, vol. 10, no. 4, pp. 3660–3667, 2018.

[153] Z. W. Shan *et al.*, "Ultrahigh stress and strain in hierarchically structured hollow nanoparticles," *Nat. Mater.*, vol. 7, no. 12, pp. 947–952, 2008.

[154] D. Kwon *et al.*, "Highly sensitive, flexible, and wearable pressure sensor based on a giant piezocapacitive effect of three-dimensional microporous elastomeric dielectric layer," *ACS Appl. Mater. Interfaces*, vol. 8, no. 26, pp. 16922–16931, 2016.

[155] S. Jung *et al.*, "Reverse-micelle-induced porous pressure-sensitive rubber for wearable human-machine interfaces," *Adv. Mater.*, vol. 26, no. 28, pp. 4825–4830, 2014.

[156] T. Yamada *et al.*, "A stretchable carbon nanotube strain sensor for human-motion detection," *Nat. Nanotechnol.*, vol. 6, no. 5, pp. 296–301, 2011.

[157] B. Wang, S. Bao, S. Vinnikova, P. Ghanta, and S. Wang, "Buckling analysis in stretchable electronics," *npj Flex. Electron.*, vol. 1, no. 1, p. 5, 2017.

[158] Y. Su *et al.*, "In-plane deformation mechanics for highly stretchable electronics," *Adv. Mater.*, vol. 29, no. 8, pp. 1–12, 2017.

[159] Y. Ma, X. Feng, J. A. Rogers, Y. Huang, and Y. Zhang, "Design and application of 'J-shaped' stress-strain behavior in stretchable electronics: A review," *Lab Chip*, vol. 17, no. 10, pp. 1689–1704, 2017.

[160] S. Gupta, W. T. Navaraj, L. Lorenzelli, and R. Dahiya, "Ultra-thin chips for high-performance flexible electronics," *npj Flex. Electron.*, vol. 2, no. 1, p. 8, 2018.

[161] F. Bossuyt, T. Vervust, and J. Vanfleteren, "Stretchable electronics technology for large area applications: Fabrication and mechanical characterization," *IEEE Trans. Components, Packag. Manuf. Technol.*, vol. 3, no. 2, pp. 229–235, 2013.

[162] T. Sekitani and T. Someya, "Stretchable organic integrated circuits for large-area electronic skin surfaces," *MRS Bull.*, vol. 37, no. 3, pp. 236–245, 2012.

[163] J. S. Noh, "Conductive elastomers for stretchable electronics, sensors and energy harvesters," *Polymers (Basel)*, vol. 8, no. 4, 2016.

[164] Y. Wei, S. Chen, X. Yuan, P. Wang, and L. Liu, "Multiscale wrinkled microstructures for piezoresistive fibers," *Adv. Funct. Mater.*, vol. 26, no. 28, pp. 5078–5085, 2016.

[165] X. Wang *et al.*, "Development of a flexible and stretchable tactile sensor array with two different structures for robotic hand application," *RSC Adv.*, vol. 7, no. 76, pp. 48461–48465, 2017.

[166] D. Qi *et al.*, "Highly stretchable gold nanobelts with sinusoidal structures for recording electrocorticograms," *Adv. Mater.*, vol. 27, no. 20, pp. 3145–3151, 2015.

[167] T. Y. Choi *et al.*, "Stretchable, transparent, and stretch-unresponsive capacitive touch sensor array with selectively patterned silver nanowires/reduced graphene oxide electrodes," *ACS Appl. Mater. Interfaces*, vol. 9, no. 21, pp. 18022–18030, 2017.

[168] S. Xu *et al.*, "Assembly of micro/nanomaterials into complex, three-dimensional architectures by compressive buckling," *Science (80-.).*, vol. 347, no. 6218, pp. 154–159, 2015.

[169] J. A. Fan *et al.*, "Fractal design concepts for stretchable electronics," *Nat. Commun.*, vol. 5, pp. 1–8, 2014.

[170] J. K. Paik, R. K. Kramer, and R. J. Wood, "Stretchable circuits and sensors for robotic origami," *IEEE Int. Conf. Intell. Robot. Syst.*, no. Sept, pp. 414–420, 2011.

[171] J. Rogers, Y. Huang, O. G. Schmidt, and D. H. Gracias, "Origami MEMS and NEMS," *MRS Bull.*, vol. 41, no. 2, pp. 123–129, 2016.

[172] L. Xu, T. C. Shyu, and N. A. Kotov, "Origami and kirigami nanocomposites," *ACS Nano*, vol. 11, no. 8, pp. 7587–7599, 2017.

[173] D. S. Gray, J. Tien, and C. S. Chen, "High-conductivity elastomeric electronics," *Adv. Mater.*, vol. 16, no. 5, pp. 393–397, 2004.

[174] M. Vosgueritchian, D. J. Lipomi, and Z. Bao, "Highly conductive and transparent PEDOT:PSS films with a fluorosurfactant for stretchable and flexible transparent electrodes," *Adv. Funct. Mater.*, vol. 22, no. 2, pp. 421–428, 2012.

[175] D.-H. Kim *et al.*, "Epidermal electronics," *Science (80-.).*, vol. 333, no. 6044, pp. 838–843 LP–843, Aug. 2011.

[176] X. Wang, H. Hu, Y. Shen, X. Zhou, and Z. Zheng, "Stretchable conductors with ultrahigh tensile strain and stable metallic conductance enabled by prestrained polyelectrolyte nanoplatforms," *Adv. Mater.*, vol. 23, no. 27, pp. 3090–3094, 2011.

[177] F. Xu and Y. Zhu, "Highly conductive and stretchable silver nanowire conductors," *Adv. Mater.*, vol. 24, no. 37, pp. 5117–5122, 2012.

[178] Y. Zhu, F. Xu, X. Wang, and Y. Zhu, "Wavy ribbons of carbon nanotubes for stretchable conductors," *Adv. Funct. Mater.*, vol. 22, no. 6, pp. 1279–1283, 2012.

[179] Y. Wang *et al.*, "A highly stretchable, transparent, and conductive polymer," *Sci. Adv.*, vol. 3, no. 3, pp. 1–11, 2017.

[180] M. A. Darabi *et al.*, "Correction to: Skin-inspired multifunctional autonomic-intrinsic conductive self-healing hydrogels with pressure sensitivity, stretchability, and 3D printability (*Advanced Materials*, (2017), 29, 31, (1700533), 10.1002/adma.201700533)," *Adv. Mater.*, vol. 30, no. 4, pp. 1–8, 2018.

[181] T. Wang *et al.*, "A self-healable, highly stretchable, and solution processable conductive polymer composite for ultrasensitive strain and pressure sensing," *Adv. Funct. Mater.*, vol. 28, no. 7, pp. 1–12, 2018.

[182] J. Li *et al.*, "Healable capacitive touch screen sensors based on transparent composite electrodes comprising silver nanowires and a furan/maleimide diels-Alder cycloaddition polymer," *ACS Nano*, vol. 8, no. 12, pp. 12874–12882, 2014.

[183] H. Jin, T. P. Huynh, and H. Haick, "Self-healable sensors based nanoparticles for detecting physiological markers via skin and breath: toward disease prevention via wearable devices," *Nano Lett.*, vol. 16, no. 7, pp. 4194–4202, 2016.

[184] T. P. Huynh and H. Haick, "Self-healing, fully functional, and multiparametric flexible sensing platform," *Adv. Mater.*, vol. 28, no. 1, pp. 138–143, 2016.

[185] D. Y. Wu, S. Meure, and D. Solomon, "Self-healing polymeric materials: A review of recent developments," *Prog. Polym. Sci.*, vol. 33, no. 5, pp. 479–522, 2008.

[186] M. D. Hager, P. Greil, C. Leyens, S. Van Der Zwaag, and U. S. Schubert, "Self-healing materials," *Adv. Mater.*, vol. 22, no. 47, pp. 5424–5430, 2010.

[187] Q. Zhang, L. Liu, C. Pan, and D. Li, "Review of recent achievements in self-healing conductive materials and their applications," *J. Mater. Sci.*, vol. 53, no. 1, pp. 27–46, 2018.

[188] Y. J. Tan, J. Wu, H. Li, and B. C. K. Tee, "Self-healing electronic materials for a smart and sustainable future," *ACS Appl. Mater. Interfaces*, vol. 10, no. 18, pp. 15331–15345, 2018.

[189] B. C. K. Tee and J. Ouyang, "Soft electronically functional polymeric composite materials for a flexible and stretchable digital future," *Adv. Mater.*, vol. 30, no. 47, p. 1802560.

[190] X. Chen *et al.*, "A thermally re-mendable cross-linked polymeric material," *Science (80–.).*, vol. 295, no. 5560, pp. 1698–1702, 2002.

[191] W. G. Skene and J.-M. P. Lehn, "Dynamers: Polyacylhydrazone revers-

ible covalent polymers, component exchange, and constitutional diversity," *Proc. Natl. Acad. Sci.*, vol. 101, no. 22, pp. 8270–8275, 2004.

[192] B. Ghosh and M. W. Urban, "Self-repairing oxetane-substituted chitosan polyurethane networks," *Science (80–.).*, vol. 323, no. MARCH, pp. 1458–1459, 2009.

[193] P. Cordier, F. Tournilhac, C. Soulié-Ziakovic, and L. Leibler, "Self-healing and thermoreversible rubber from supramolecular assembly," *Nature*, vol. 451, no. 7181, pp. 977–980, Feb. 2008.

[194] Y. Chen, A. M. Kushner, G. A. Williams, and Z. Guan, "Multiphase design of autonomic self-healing thermoplastic elastomers," *Nat. Chem.*, vol. 4, no. 6, pp. 467–472, 2012.

[195] C. Wang *et al.*, "A rapid and efficient self-healing thermo-reversible elastomer crosslinked with graphene oxide," *Adv. Mater.*, vol. 25, no. 40, pp. 5785–5790, 2013.

[196] N. Holten-Andersen, M. Harrington, H. Birkedal, *et al.* "pH-induced metal-ligand cross-links inspired by mussel yield self-healing polymer networks with near-covalent elastic moduli". PNAS, vol. 108, no. 7, pp. 2651–2655, 2011.

[197] M. Burnworth, L. Tang, J. Kumpfer, *et al.* "Optically healable supra-molecular polymers". *Nature*, vol. 472, no. 7343, pp. 334–337, 2011.

[198] M. Nakahata, Y. Takashima, H. Yamaguchi, and A. Harada, "Redox-responsive self-healing materials formed from host-guest polymers," *Nat. Commun.*, vol. 2, no. 1, pp. 511–516, 2011.

[199] D. C. Tuncaboylu, A. Argun, M. Sahin, M. Sari, and O. Okay, "Structure optimization of self-healing hydrogels formed via hydrophobic interactions," *Polymer (Guildf).*, vol. 53, no. 24, pp. 5513–5522, Nov. 2012.

[200] U. Gulyuz and O. Okay, "Self-healing poly(acrylic acid) hydrogels with shape memory behavior of high mechanical strength," *Macromolecules*, vol. 47, no. 19, pp. 6889–6899, 2014.

[201] Q. Wang *et al.*, "High-water-content mouldable hydrogels by mixing clay and a dendritic molecular binder," *Nature*, vol. 463, no. 7279, pp. 339–343, 2010.

[202] K. Haraguchi, "Stimuli-responsive nanocomposite gels," *Colloid Polym. Sci.*, vol. 289, no. 5–6, pp. 455–473, 2011.

[203] J. Y. Sun *et al.*, "Highly stretchable and tough hydrogels," *Nature*, vol. 489, no. 7414, pp. 133–136, 2012.

[204] T. L. Sun *et al.*, "Physical hydrogels composed of polyampholytes demonstrate high toughness and viscoelasticity," *Nat. Mater.*, vol. 12, no. 10, pp. 932–937, 2013.

[205] A. Miyamoto *et al.*, "Inflammation-free, gas-permeable, lightweight, stretchable on-skin electronics with nanomeshes," *Nat. Nanotechnol.*, vol. 12, no. 9, pp. 907–913, 2017.

[206] S. Choi *et al.*, "Highly conductive, stretchable and biocompatible Ag–Au core–sheath nanowire composite for wearable and implantable bioelectronics," *Nat. Nanotechnol.*, vol. 13, no. 11, pp. 1048–1056, 2018.

[207] B. W. An, S. Heo, S. Ji, F. Bien, and J. U. Park, "Transparent and flexible fingerprint sensor array with multiplexed detection of tactile pressure and skin temperature," *Nat. Commun.*, vol. 9, no. 1, pp. 1–10, 2018.

[208] Y. Shi *et al.*, "Electronic synapses made of layered two-dimensional materials," *Nat. Electron.*, vol. 1, no. 8, pp. 458–465, 2018.

[209] Y. van de Burgt, A. Melianas, S. T. Keene, G. Malliaras, and A. Salleo, "Organic electronics for neuromorphic computing," *Nat. Electron.*, vol. 1, no. 7, pp. 386–397, 2018.

[210] C. Bartolozzi *et al.*, "Neuromorphic Systems," *Wiley Encycl. Electr. Electron. Eng.*, pp. 1–22, 2016.

[211] L. Y. Chen *et al.*, "Continuous wireless pressure monitoring and mapping with ultra-small passive sensors for health monitoring and critical care," *Nat. Commun.*, vol. 5, pp. 1–10, 2014.

[212] S. Han *et al.*, "Battery-free, wireless sensors for full-body pressure and temperature mapping," *Sci. Transl. Med.*, vol. 10, no. 435, 2018.

[213] A. M. Hussain, F. A. Ghaffar, S. I. Park, J. A. Rogers, A. Shamim, and M. M. Hussain, "Metal/Polymer based stretchable antenna for constant frequency far-field communication in wearable electronics," *Adv. Funct. Mater.*, vol. 25, no. 42, pp. 6565–6575, 2015.

Acknowledgments

B. C. K. T. is grateful for the support of the National Research Foundation Fellowship (NRFF-2017-13) from the National Research Foundation of Singapore, Prime Minister Office, and the National University of Singapore (NUS) Presidential Young Professorship. H. Y. acknowledges support from the NUS Research Scholarship. M. H. acknowledges support from the NUS National Graduate Scholarship (NGS).

Cambridge Elements ≡

Flexible and Large-Area Electronics

Ravinder Dahiya
University of Glasgow

Ravinder Dahiya is Professor of Electronic and Nanoengineering, and an EPSRC Fellow, at the University of Glasgow. He is a Distinguished Lecturer of the IEEE Sensors Council, and serves on the Editorial Boards of the *Scientific Reports*, *IEEE Sensors Journal* and *IEEE Transactions on Robotics*. He is an expert in the field of flexible and bendable electronics and electronic skin.

Luigi G. Occhipinti
University of Cambridge

Luigi G. Occhipinti is Director of Research at the University of Cambridge, Engineering Department, and Deputy Director and COO of the Cambridge Graphene Centre. He is Founder and CEO at Cambridge Innovation Technologies Consulting Limited, providing research and innovation within both the health care and medical fields. He is a recognised expert in printed, organic, and large-area electronics and integrated smart systems with over 20 years' experience in the semiconductor industry, and is a former R&D Senior Group Manager and Programs Director at STMicroelectronics.

About the Series

This innovative series provides authoritative coverage of the state of the art in bendable and large-area electronics. Specific Elements provide in-depth coverage of key technologies, materials and techniques for the design and manufacturing of flexible electronic circuits and systems, as well as cutting-edge insights into emerging real-world applications. This series is a dynamic reference resource for graduate students, researchers, and practitioners in electrical engineering, physics, chemistry and materials.